Outer
Space

Outer
Space

PROSPECTS FOR
MAN & SOCIETY

♦ The American Assembly
Columbia University

A SPECTRUM BOOK

Englewood Cliffs, N. J.

PRENTICE-HALL, INC.

Preface

The object of this book is to render intelligible the complex issues and alternatives which confront us on the fantastic frontier beyond the earth. This original material, written under the supervision of Lincoln P. Bloomfield of the Massachusetts Institute of Technology, first provided the background for the Twentieth American Assembly at Arden House, the Harriman Campus of Columbia University. The result of those deliberations begins on page 193.

Across the United States and abroad, other Assembly meetings as well as individual readers may ponder the challenges of the unprecedented space dimension defined in these chapters and come to their own conclusions. This non-partisan venture in space education, supported by the Ford Foundation, represents no position of the Foundation or of The American Assembly.

Henry M. Wriston
President
The American Assembly

Table of Contents

Outer
Space

Introduction:

The space revolution—
a perspective

◆ Lincoln P.
Bloomfield

Editor

Seen through the long lens of history only a few scientific and technical events have the stature of genuine revolutions in human affairs. The Promethean discovery of fire, the first wheel, the invention of gunpowder, the printing press, the germinal notions of Copernicus, Galileo and Newton, cheap electricity, steam power, and, in our time, $e = mc^2$—these are the great landmarks of the physical transformations in the way men have lived, thought, and acted.

Twentieth century man, still grappling with the after-effects of the industrial revolution and barely able to discern the ultimate meaning in the released energy of the atomic nucleus, must now lift up his eyes to the heavens—the very seat of the mysteries that envelop his life and

1

♦ LINCOLN P. BLOOMFIELD is Director of the Arms Control Project at the Center for International Studies of the Massachusetts Institute of Technology. As Associate Professor of Political Science he also initiated MIT's teaching and research programs on the political implications of outer space. Prior to coming to MIT in 1957 to direct the Center's United Nations project, he served for eleven years in the State Department, most recently as the Department's policy planner on UN Affairs. Now consultant to various government agencies and lecturer at the Army, Navy and Air War Colleges, he is author of *The United Nations and US Foreign Policy*, and *Evolution or Revolution?—The UN and the Problem of Peaceful Territorial Change*.

mark his place in the cosmos. Somehow, he must assimiliate yet another transforming revolution, this time in the dimension that has been most closed to him, most forbidding, most elusive—and most tantalizing in its unrevealed promises.

In the end, space will affect everyone. Already the lives of millions of Americans and Russians are affected indirectly as they provide the staggering sums of money to finance space activities. Human beings everywhere will be even more directly affected if the military nightmare comes true of multi-megaton weapons, carried by earth-circling satellites, deliverable to the ground below by a coded radio command. Yet thus far in its brief life-span outer space, more than any other of the lengthening reaches of contemporary activity, has belonged, so to speak, to the few who knew its language. This revolution, because of its complex and specialized nature, is essentially still in the hands of a small èlite whose knowledge, and judgments about that knowledge, have so far been only feebly communicated to mankind at large.

This will not do. It will not do because of the effects of space, present or potential, on every man. Above all, it will not do because with Thomas Jefferson we still

> . . . know of no safe depository of the ultimate powers of society but the people themselves; and if we think them not enlightened enough to exercise their control with a wholesome discretion, the remedy is not to take it from them, but to inform their discretion by education.

Education of the layman is the prime aim of this book, as it is of the whole enterprise of The American Assembly. The Congressman has to appropriate billions for space—or have good reasons for refusing to do so. Businessmen and investors need to make decisions which their experts can only advise them about. Ultimately the voter and taxpayer, in accordance with the Jeffersonian theorem, makes the most basic decision of all and needs help in doing so.

Thus this volume seeks to communicate in everyday language the implications of the new technology for life upon earth, particularly the impact to be anticipated in our familiar economic, social, and political environments. To the extent the effort is successful it may open up still further the needful dialogue between the expert and the sovereign people. For while the experts may and do propose, the people, in the last analysis, will dispose. The layman, even if he cannot become an expert, can become informed. Indeed, he has no choice if he is to act responsibly when lay judgments about technical matters are placed before him.

Those with special knowledge have their duty also. They must solve for themselves the problems of balance, of perspective, of political realism and social meaning. It is not that we need fear the nightmare of technocratic rule—Plato's philosopher-kings with engineering degrees, as it were. But it is equally urgent that our unfolding technical knowledge be civilized with human perspectives. As Archibald MacLeish recently wrote,

> . . . the knowledge of the fact has somehow or other come loose from the *feel* of the fact, and . . . it is now possible, for the first time in human history, to know as a mind what you cannot comprehend as a man. . . . Not until mankind is able again to see *feelingly* . . . will the crucial flaw at the heart of our civilization be healed.

Another purpose of this collection of essays, then, is to help us all to see the age of space feelingly, and intelligibly in terms of both, in Sir Charles Snow's word, the cultures. Lamentably, the two cultures are still far from symmetrical. It is surely trite to cite the lag between scientific technology and the human capacity to maximize its benefits and minimize its harm through social, economic and political arrangements. Still, little out of contemporary social science can yet match with assurance physical science's quantum leap into outer space. Dr. Karl Menninger speaks for more than the psychiatrists in finding it "disturbing to compare the advances in the physical sciences that make such flight into space possible with our pedestrian progress in psychology and the behavioral sciences."

But the space revolution has no ground rules that provide for time-out while the behavioral sciences catch up. Neither the steam engine nor the atomic engine waited; the assault upon space is not waiting. The earth has been orbited by scores of satellite vehicles, several containing men; the sun has been circumnavigated; the moon has been hit, and a mechanical messenger from earth reported back until it had reached a distance of 22 million miles out. Communications, weather forecasting, navigation, transportation—none of these will ever be the same again. With few apparent inhibitions of an economic nature the Soviet Union and the United States are competing in a race to place their nationals on the surface of the moon and, not long afterward, on the nearer and presumably more hospitable planets. More ominously, the technology of space is already invaded with the same virus that infects life on earth—the capacity, now or in the near future, for ever more efficient means of annihilating human life on a grand scale, not to mention espionage, hostile propaganda, and threats designed to break another's national will.

So while we can and should continue to argue the merits and the means of the competition in space, it is truly upon us. Whatever the details of execution, the process to which the great powers have now committed themselves appears to be irreversible. It is the Communists who have defined the terms of the contest so far—spectacular manned moon and planetary probes requiring huge amounts of booster thrust. The United States resisted the prestige race for several years on grounds of economy, of lack of military requirements, and of preference for other valid scientific objectives in space. But any doubt that the United States would finally face up to the broader challenge was dispelled on May 25, 1961, when President Kennedy urged that this nation "commit itself to achieving the goal, before this decade is out, of landing a man on the moon and returning him safely to the earth." The President estimated that such a program is going to cost 7 to 9 billions of dollars over the next five years; the reader will find in this volume aggregated figures (including military costs) for the decade ranging up to the staggering total of 100 billions.

"This decision," warned the President, "demands a major national commitment of scientific and technical manpower, material and facilities, and the possibility of their diversion from other important activities where they are already thinly spread." But he warned, "while we cannot guarantee that we shall one day be first, any failure to make this effort will make us last." The time has arrived "for this nation to take a clearly leading role in space achievement which, in many ways, may hold the key to our future on earth."

Some crucial preliminary choices, then, have been made. But the debate is by no means over. A Gallup poll in May 1961 reported that only 33 per cent of those Americans polled favored spending $40 billions to get a man to the moon, and 58 per cent were opposed. The question of ultimate public support remains before us, along with virtually all the grand issues of public policy. This volume seeks to illuminate some of them.

Obviously, not all the issues could be covered here—but those at hand make a heavy agenda. First of all, what are the facts about the "hardware"? If the lay citizen once was able to shy away from technical facts because they were irrelevant to his concerns, with such an attitude today he will surely fail in his duty. The need to make responsible political judgments about technical matters cannot be deferred until our educational system produces a new race of intellectual hybrids. H. Guyford Stever's chapter on the technical considerations bearing on outer space is presented first, without apologies, and with the hope that it will be read first. One is not entitled to believe that the necessary technical briefing will always be as painless as Dr. Stever has made it here.

The next three chapters involve primarily the nation and its welfare. What are the concrete benefits that can actually be expected from space? Donald Michael enumerates them and as a social psychologist analyzes their feedback to the earthly environment. Leonard S. Silk, writing as both an economist and a business analyst, then suggests the incredible diversity and proliferation of the industrial effort that has been—and will be—involved. What shall be our philosophy of administration, and how reconcile the sometimes conflicting claims of the public and private sectors? T. Keith Glennan, with incomparable experience as the first administrator of the United States space program, suggests some benchmarks.

The next three chapters reach out into the world of nations. What can the scientists do together, across national boundaries, to deal with phenomena which recognize no political barriers? Hugh Odishaw, out of his unique role in international scientific cooperation, reminds us of the part the individual can still play in a sometimes inhuman world. Then Donald Brennan draws comprehensively on two complementary professional skills to outline the two faces of space, seen as a military problem: weapons, and their control. Thirdly, I myself have tried to lay out some of the uncommonly baffling issues of law and politics that connect the realm of outer space with the world of diplomacy and power.

Finally, the question with which the dialogue should both begin and end: is the allocation of resources the nation has begun to set for itself a rational one, in balance with other national goals and objectives? James R. Killian, Jr., considers this question against the background of his role of leadership in the nation's total scientific effort.

*   *   *

In the end, the problem of outer space is a problem of human values. The values of scientific discovery, of exploration of the unknown, of convenience, of rapidity—all of these have genuine meaning only in reference to the human being and his society. If we are to be faithful to the humanistic traditions behind us, we must ask, if I may paraphrase President Kennedy, not only what we can do in space, but what space will do to us.

Yet at the same time there is a vital place in our scheme of values for the quest of the unknown and the hitherto unknowable. The conquest of space has all the elements of a great historic drama, and life without drama is dull indeed. Above all, and in the highest creative sense, it is an opportunity for statesmen to build their political structures in a still relatively uncluttered area of interaction between the nations. To be sure, there is nothing in the record so far to guarantee that man is capable of transcending in space the conflicts, which have kept his earthly home in turmoil and peril. All we can do is hope that the ever-accelerating thrust into this new realm will in turn push social invention to the point where it has a chance of catching up in the race of history.

Whether at home, in formulating our national space policies, or in seeking to construct a better design for managing men's affairs in the world at large, the task lies ready at hand. Reinhold Niebuhr, with his customary wisdom, supplied the relevant perspective when he wrote:

> It is man's ineluctable fate to work on tasks which he cannot complete in his brief span of years, to accept responsibilities the true ends of which he cannot fulfill, and to build communities which cannot realize the perfection of his visions.

1.

The technical prospects

◆ H. GUYFORD STEVER

A serious prognosis of the technical prospects of space flight requires more than a mercurial judgment about the quick attainment of some of the projects now being discussed. It requires an appreciation of the history of technology, of how new technologies unfold. There is a striking parallel between the history of the airplane and the history of space flight to date. A review of this parallel can show the kinds of indicators to be looked for in estimating the prospects for the future of space flight. (No attempt is made here to detail the history of the airplane. For those interested, reference is made to *The Airplane,* a superb historical survey by Charles H. Gibbs-Smith.)

THE ANALOGY OF THE AIRPLANE

Man's early dreams of flight in the atmosphere like a bird were intermingled with his

♦ H. GUYFORD STEVER is Professor of Aeronautics and Astronautics at the Massachusetts Institute of Technology. One-time Chief Scientist of the United States Air Force, he is Vice Chairman of the Air Force Scientific Advisory Board and Chairman of NASA's Research Advisory Committee on Missile and Space Vehicle Aerodynamics. In 1960 he was elected President of the Institute of Aerospace Sciences.

dreams of space flight to other planets throughout the solar system. Confusion existed because natural philosophers did not have an accurate picture of the extent of the atmosphere. The early myths of manned atmospheric and space flight include that of Daedalus, builder of the Minoan Labyrinth, and his son Icarus, that of King Bladud, tenth legendary King of England and father of King Lear, and a host of other tales, some of which may well be based on early, probably tragic, experiments in flying with bird-like wing constructions. As science began to develop in the Middle Ages, more practical dreamers, if we may use such a term, such as the Franciscan monk, Roger Bacon, and later Leonardo da Vinci, had ideas which might have led to practical embodiment had the technology been sufficiently advanced. For over two hundred years before flight was achieved, physical experiment and attempts to fly increased steadily, and more and more men began to get the concept of powered flight.

The engineering basis of flight was laid in the early nineteenth century, long before the Wright Brothers first flew, by Sir George Cayley, a British minor nobleman, who was a brilliant engineer in many different fields. His accomplishments in aeronautics, though not widely appreciated, were astounding. He was the first to realize that the airplane would attain the lift needed to counteract its weight by a thrusting device including a propeller and an engine which would overcome the drag of the air. Among his many other aeronautical accomplishments he designed and built the first model of a practical airplane. But, more important, he had a very clear vision of the future and in 1809, almost a century before manned controllable powered flight was achieved, he wrote:

> I may be expediting the attainment of an object that will in time be found of great importance to mankind, so much so that a new era in society will commence from the moment that aerial navigation is familiarly realized—I feel perfectly confident, however, that this noble art will soon be brought home to man's convenience and that we shall be able to transport ourselves and our families and their goods and chattles more securely by air than by water and with velocities of from 20 to 100 m.p.h.

Though his numbers fell short of the mark, he had the spirit of the modern development of air transportation. These were the words of an imaginative but still practical engineer.

On the other hand, few basic research scientists had anything to do with the attainment of flight, nor were they sanguine about its use. They generally ignored the field or discounted it. For example, Lord Kelvin, one of the world's great research physicists, said in 1896, only seven years before the attainment of controllable manned powered flight, "I have not the smallest molecule of faith in aerial navigation other than ballooning."

Even after the achievement of controllable powered-manned flight and after many people had followed the Wright Brothers' lead, it was still difficult to foresee the future. In 1908 the Wright Brothers, who were still leading in the development of airplanes all over the world, delivered to the United States Army an airplane to fulfill a contract which called for a flight speed of about 32 m.p.h. It was constructed of airplane cloth and a hickory wood frame; it had two small 9-foot propellers geared by belt drives to a single motor of which the power output was about 25 horsepower, lower than that of practically any modern automobile. You recall the pictures of the Wright Brothers' Flyer, with its fixed horizontal tails in the front, vertical rudders to the rear. It did not even have wheels—just skids. It was normally launched with a catapult mechanism, though occasionally Wilbur Wright was skillful enough to take it off on wet grass. A standard stunt in those days was for a man or two to push on the rear of the wings to help the airplane get started.

Still the concept of flight was exciting to enough people so that in its infancy many predictions were made of the technical prospects of the airplane and of its use to mankind. Not all were imaginative. In 1910, for example, the British Secretary of State for War said, "We do not consider that airplanes will be of any possible use for war purposes."

The first uses of the airplane which spurred its development were military. It was an improved means of performing certain limited military tasks such as observing the enemy. To most minds its function was to replace the cavalry as the eyes of the army and the balloon as the spotter for the artillery.

It was very difficult for a practical man in 1909, when the first Wright planes were being adopted for military use, to conceive of commercial air transportation. Although calculations of the air-transportation economics of those days vary quite considerably, they point up some basic facts. In 1909 a plane could travel at 42½ m.p.h. with a pilot and a single passenger. The plane had a useful life of only a small number of hours— possibly 30—and it cost $30,000. The cost per passenger mile might then turn out to be something like $25 per passenger mile or, in 1960 dollars, $80 per passenger mile. Today the operating cost of a jet airplane which flies more than ten times as fast and has a useful range of almost

100 times as great with 100 times as many passengers is only a few cents per passenger mile.

No one accurately foresaw the shape of things to come for the airplane. Those who had faith that technology has a future came closest to predicting the future. There is a story, possibly aprocryphal, that the head of the Astor business enterprises said that important men would conduct their business by traveling in airplanes in 50 years. He did not worry about the limits to the load-carrying capacity of wood and fabric airplanes. He did not worry that there were limits in the power available. He did not worry that flying an airplane at that stage was dangerous. He did not even stop to consider the tremendous development cost.

Astor went right to a useful purpose that the knowledge of the day promised, and his faith in technology proved right. The structures changed from cloth and wood to metal. Steel and the light aluminum and magnesium alloys were developed, and the technique of stressing the skin instead of using a bracing framework brought aeronautics to its modern era. The engines developed from a few tens to many thousands of horsepower; internal combustion gasoline engines with propellers were replaced by turbojet engines. The vast technological improvements in every field of engineering associated with the airplane have made commercial flight commonplace.

The parallel between the story of the airplane and that of the applications of our space technology is obvious. In the military context, the first space concepts were observation satellites. The achievement of a bombardment capability from space and space combat is now being given serious thought and development. Eventually military operations may well be conducted simply for control of space as in the past they have been conducted for control of the air. In the context of peaceful applications, there has been some slower development.

Only after major emphasis on military uses do we now appear to have within our reach world-wide communications by satellite relay stations, a world-wide weather observation and prediction service using satellites, and a world-wide navigation system for ships using navigation satellites.

THE BASIS FOR PREDICTIONS

The lessons in prognosticating the technical possibilities of space flight that can be learned from this brief consideration of another great technology are numerous. For example, most people, over the long run, fall short of the mark in their predictions. Developments follow the lines of practical use. Military developments lead the way to nonmilitary applications.

I have history in mind, then, as I attempt here to look ahead to the technical prospects for outer space. Moreover, I have in mind certain very present factors which bear on any forecast.

In the latter half of the nineteenth century, when classical science was flourishing, J. Henri Poincaré wrote in *La Science et l'Hypothèse*: "For a superficial observer, scientific truth is beyond the reaches of doubt; scientific logic is infallible and, if scientists sometimes err, it is because they have misunderstood the rules." [1] If the task of presenting the technical prospects for outer space depended only upon understanding scientific truths, the future could be plotted with reasonable simplicity and confidence for some time ahead. But progress in space will not be essentially or solely scientific. It will involve engineering, in which the laws of science play an important but only partial role. Thus progress in space, like that in all engineering projects, will be critically affected by economic and social factors.

For decades space progress can be made by practicing in new and generally more expensive embodiments the arts we already know. It will depend upon engineers who must improve the design of existing equipment, design similar equipment in larger sizes, and develop new devices in fields of engineering where the principles are well known. We can already identify some of the areas in which those developments will be made. Steady but not overwhelming gains can be made in liquid and solid propellant rocketry. Nuclear rocketry and electrical particle rocketry are being developed with the promise of vast improvement in space capability. Some of the most important but least publicized gains in the recent past and expected gains in the near future are in the fields of structural design and materials. Auxiliary power is a key field of future development. Communications, radio and inertial guidance and other space navigation developments will be needed before useful space accomplishments can unfold in large number. The engineering of life-support equipment for human flight is a relatively new field which offers great promise of improvement.

It is clear also that space progress will depend upon the financial support given to the development organizations of which we already have many more in this country than we are using efficiently. Moreover, it will depend upon the size of the continuing military effort.

However, any prediction based solely on our current technology, with reasonable estimates of government interest and financial support, would most certainly lead to an underestimate of the technical prospects for outer space. As any student is aware, future progress in engineering will depend upon developments not known in today's art; and the talented young men and women who are now going through training in engineering and science, much better equipped than earlier generations both in background and in their approach to education, will march to the future of technology more rapidly than we now estimate. One can be sure that there will be major new developments which are not foreseen today, and that some of them will move the space program forward faster and farther than we can predict.

---

[1] Translated from the French.

At the outset of this chapter, then, I want to declare that I am an enthusiast for the long-term potential of space flight. I can best describe my attitude by telling an anecdote about a foreign visitor who took a taxi tour of our national capital. When shown the Archives building, on which there is inscribed a quotation from Shakespeare's "The Tempest" which reads "What is past is prologue," the foreign visitor was a little puzzled, for he did not have a good command of the English language. He asked his taxi driver if he knew what the saying meant. The taxi driver answered, "Sure, bud, that means you ain't seen nuthin' yet."

I believe that we have only scratched the surface of the technology of space flight. I believe that we have the trained engineers in the aerospace field with enthusiasm and vision who can achieve their promises.

Space Flight Velocity Requirements

The velocity requirements for various space missions have nothing to do with the past, present, or future state of technology. They come out of a very old branch of science, celestial mechanics, which began with the ancients as they studied the motion of the stars, was given a big boost by Copernicus and mathematical foundation by Kepler and Newton, and grew to its peak many decades ago as astronomers made the system accurate. In fact, it was a science which was almost in mothballs until the new-found rocket technology returned it to prominence.

THE PROBLEM OF PROPULSION

The key technology in space flight is propulsion. Rocket boosters are now capable of accelerating useful payloads to the very high velocities which are required to orbit the Earth, to escape the Earth and go to the Moon and the other planets, and to orbit the Sun. Most alert readers of the newspapers in recent years have amassed a few characteristic numbers which describe the high velocities required for space flight. For the purposes of this chapter, in describing the speeds it is worth introducing an illustration (see graph: *Velocity Requirements,* etc.). Incidentally, this graph could have been prepared by Sir Isaac Newton using his newly enunciated Law of Universal Gravitation and the mathematical tools available to him.

The graph shows the velocity required for a body to move on an elliptical orbit starting and terminating on the surface of the Earth as the ballistic missile does, in a circular satellite orbit around the Earth as a Sputnik, a typical Earth satellite, or as the Moon does, and in an elliptical orbit changing from the Earth's orbit around the Sun to one of the planet's orbits around the Sun. Though Sir Isaac would have been capable of plotting such a graph, he certainly would have objected to the entire proceeding as

VELOCITY REQUIREMENTS FOR BALLISTIC MISSILE AND SPACE FLIGHT

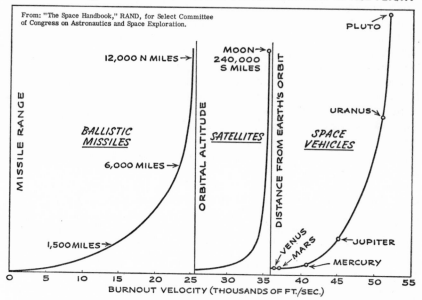

From: "The Space Handbook," RAND, for Select Committee of Congress on Astronautics and Space Exploration.

not being scientific since only a dreamer would think that man could ever have the capability of attaining such velocities in any useful vehicle.

Velocities are given in feet-per-second. Many readers are more familiar with velocities given in miles-per-hour. It is easy to convert approximately from feet-per-second to miles-per-hour by taking two-thirds of the number. For example, 15,000 ft/sec is roughly 10,000 m.p.h.

First, consider the speeds required for ballistic missiles. The very high speed of 5,000 ft/sec, about one mile/sec or 3,600 m.p.h., enables a ballistic missile to achieve a range somewhere between 200 and 250 miles. An increase in speed to 15,000 ft/sec, on the other hand, enables the vehicle to attain a 1500-mile range. A speed of less than double that—between 23- and 24,000 ft/sec—permits a factor of four increase in range from 1500 to 6000 miles. An increase in speed from about 23,500 to 25,000 ft/sec increases the range from 6000 to 12,000 miles; and a very slight increase of velocity over the 25,000 ft/sec puts a satellite into orbit at low altitude.

Further increases in velocity capability from 26,000 feet to about 36,000 ft/sec permit the satellite orbit to be established at increasingly high altitudes to a point where at something over 36,000 ft/sec the gravitational pull of the Earth can be entirely escaped so that the vehicle would then be in orbit around the Sun just as the Earth is. With a few thousand ft/sec more in speed, a space vehicle can get to the regions of Mars and Venus, say at 41,000 ft/sec, to Mercury with 45,000 ft/sec, Jupiter with 51,000 ft/sec, and so on.

These speed requirements are well known to space engineers; in fact all of them have these numbers at the tip of their fingers at all times and, in this era of advertising publicity and public speeches, they are not only at the tip of the space engineer's fingers but also at the tip of his tongue.

The speeds given on the chart are minimum to achieve the objectives. If the mission requires some special maneuvering such as landing on a planet, the speeds are somewhat higher. For example, if one is considering a round-trip lunar flight in which the vehicle takes off from the Earth, uses a rocket to brake its speed as it decelerates to land safely on the Moon, takes off from the Moon, and comes back to the Earth using atmospheric braking here on the Earth, the speed capability of the rocket should not be just 36,000 ft/sec, but something like 60,000 ft/sec. Likewise, if a spaceship is required to go from Earth to Mars on the minimum velocity of about 37,000 ft/sec, the spaceship can do this only when Mars is in the optimum position and it could not use any rocket braking or other maneuvers around Mars. On the other hand, with a capability of 100,000 ft/sec change due to rocket thrust, instead of just the minimum 37,000 ft/sec speed capability, it could propel itself there in 15 or 20 days; if it had 300,000 ft/sec it could make the trip in 10 days. So the figures given on the graph are misleading with respect to space missions of an advanced nature. In reality one would like to be able to design very high speed increments into the rocket boosters which enable the space vehicle to perform its mission.

SOME COMPARISONS

One might digress here in order to put these very large velocities into context with other high velocity devices that are well known. The long history of ballistics and firearms has led to developments in which small-arms can now have velocities from 1000 ft/sec to between 2- and 3,000 ft/sec. Certain very high performance guns can go to higher velocities, and in the laboratories for special research purposes there are gun-type devices which go to 10,000 ft/sec and more. Jules Verne in describing his imaginary trip to the Moon employed a very long gun barrel with a very special new explosive to propel his ship to the Moon. Even if one could use some kind of gun-like projector for a ship for space flight, it would have several drawbacks. The first of these is that, since the highest velocity is attained right at the end of the gun barrel, which presumably would be within the atmosphere, all the difficulties of high-velocity frictional heating would plague the vehicle during takeoff. Moreover, the velocity loss in the atmosphere due to drag would be very large. In addition, there would be immense problems of high acceleration loading (high "G's") on the vehicle because the full acceleration would take place in the very short gun barrel. For these reasons the rocket principle is used.

Rockets have the tremendous advantage that the accelerations are least in the beginning and stay relatively small, small enough to be withstood by humans and by delicate instruments. The very high velocities can be reached because the accelerations occur over long periods. Furthermore, the extreme velocities are not reached until the denser portions of the atmosphere are cleared by the vehicle.

Boosters

ROCKET BOOSTER TECHNOLOGY

The concept of using rockets to attain the very high velocities for space flight is rather old. In fact, it would be difficult to pinpoint accurately the first man to conceive this. In the nineteenth century a Russian, Tsiolkowsky, a minor schoolteacher, discussed rocket power as the means by which the high velocities required for space flight could be reached. A German named Ganschwindt independently did the same.

Robert Goddard, an American physicist who started thinking along these lines during World War I, also discussed and placed on a much more scientific basis the calculations for rocket propulsion needed for space flights. He spent his whole professional career in efforts which initiated modern liquid propellant rocket technology and designed vehicles which were the forerunners of today's space vehicles, reaching in 1926 the point at which his first propellant rocket vehicle was fired.

Most of the achievements in space flight have been made using the liquid propellant rockets which were pioneered by Dr. Goddard, developed to a reasonably high state of the art by Germans in their research and development leading to the V-2 and other rocket weapons used in World War II, and developed further in both Russia and the United States mostly for ballistic missiles and only lately for spacecraft. The modern interest in rocketry revived a much more ancient type, solid propellant rocketry, started by the Chinese in the twelfth century and used sporadically but relatively ineffectively in warfare from that time until World War II, when a large number of rockets using solid propellants, such as anti-tank air-to-ground rockets, anti-submarine rockets, and bazooka rockets for infantry against tanks, became quite effective weapons. In the period following World War II solid propellant rockets also have been developed to a point where they are now figuring in current and future space plans. Though their performance is not yet quite up to that of the liquid propellant rockets, this is partially compensated for by their higher reliability and greater simplicity in operation.

Where do we stand with respect to the speed increment that can be given to a vehicle as it is shot off into space? Not only can we fire ballistic missiles more than a quarter of the way around the world; we have established circular satellites around the Earth and sent vehicles toward the Moon. Beyond the Moon, vehicles have escaped the gravitational pull of the Earth to pass near Venus, be captured in the gravitational pull of the Sun, and remain permanent satellites of the Sun. According to the chart this means that we have attained velocities in the region of 40,000 ft/sec. This represents quite an advance in speed capability when we recall that in World War II, when the V-2 was put into operation, the best speed was a little more than 5,000 ft/sec and that, only 34 years ago, Goddard's rocket got only to 184 feet in altitude.

Too often, in considering the advances made in space boosters over the recent decades, exaggerated emphasis is laid on the rocket engine. The improvement of the performance of the rocket engine is only part of the story. An important part has to do with the improvement in vehicle design in which the relative weight of the vehicle components has constantly been reduced; and there is also the final element of design to obtain the very high velocities desired, that is, multi-staging. All very high-velocity space vehicles and even ballistic missiles—the longer-range ones—are boosted by multiple stage rockets. The principle of this staging is very simple: if a single stage rocket can, say, boost a payload to half of the velocity required for a given mission, then the full velocity can be achieved by adding a larger booster stage. This booster stands in weight ratio in the same relationship to the original rocket plus payload as does the original rocket—which now becomes the second stage—in relation to the payload. This staging device has the advantage of enabling the high speeds required for the mission to be obtained; it has the tremendous disadvantage that the multiplying factor mentioned goes up very rapidly.

Suppose, for example, a mission of 25,000 ft/sec is considered. If a rocket can be designed to push a payload of 1,000 pounds to 12,500 ft/sec with the total rocket weight being ten times the weight of the payload, or 10,000 pounds, then the full velocity for the mission—the 25,000 ft/sec—can be achieved by taking the 10,000 pounds of the first rocket, and with the same ratio of ten times for a larger booster stage, or 100,000 pounds, the 100-pound payload can be boosted to 25,000 ft/sec. Carrying this same reasoning a little farther, if the mission calls for 37,500 ft/sec which would permit it to escape the gravitational pull of the Earth, the 100,000 pound total vehicle would again have to have a still larger booster stage added which was ten times its weight—or a million pounds. So the staging principle allows one tenth by weight of the first stage rocket to be boosted to 12,500 ft/sec or one one-hundredth by weight of a two-stage rocket to twice that speed or 25,000 ft/sec, or one one-thousandth by weight of a three-stage rocket to a speed of 37,500 ft/sec. One can carry on the arithmetic from there and see that it gets both expensive and discouraging to increase the stages far beyond three or four.

Since the days of the V-2, when the single-stage velocity was of the order of 5,000 ft/sec, improvements in rocket efficiency and in structural efficiency have made it possible for a single stage to reach 15 to 20,000 ft/sec. No single-stage rocket has yet reached the 25,000 ft/sec required for orbiting the Earth, though vehicles which are almost single-stage vehicles—like the Atlas, which instead of dropping off the stage only drops off some excess rocket engines, and is therefore called a one-and-one-half stage vehicle—have reached this velocity of orbiting. With today's technology one thinks of one or two stages for long-range ballistic missiles, two or three stages for orbiting vehicles, and three, four, five, or six for vehicles to go to the Moon and to escape the Earth's gravitational pull— to go to Venus or Mars or just to become ordinary satellites of the Sun. Research vehicles have been used with as many as seven stages.

One may ask the question: Can there be a radical increase in the speed increment which is obtainable from a single stage of a booster? As indicated before, such an increase must come from improving the efficiency of the engine itself or from the improvement in the structural efficiency.

Let us first look at the efficiency of the rocket itself. Over the past period of development the rocket motor design has been given a tremendous amount of attention, but for a given rocket propellant such as liquid oxygen and kerosene the expected improvement in rocket motor design cannot be very great. A given engine may be made somewhat more efficient with long and expensive development programs on the turbine fuel pumps, on the inlet design, on the jacket cooling, on the materials used, and so on. But only small gains can be made. Larger gains can be made by changing rocket propellants completely, and steps have been taken along this line. The standard rocket propellants were liquid oxygen and kerosene for the very long range ballistic missiles and spacecraft of the recent past. More recently liquid oxygen-liquid hydrogen engines have been developed with higher performance figures. There are other possible improvements using liquid hydrogen-liquid fluorine and so on. In the solid propellant field there are also possible new propellant combinations that can be made, and such tricks as making a combined liquid and solid propellant rocket are under development. One can expect some improvements then, in the propellant efficiency, but they will come in small increments and only following long-term and very expensive projects.

Considering the possibility of improving the velocity increment obtainable by a single stage by increased structural efficiency, one should point out that from the days of the German V-2 only about 70 per cent of the booster was rocket fuel. Today engineers have been achieving almost 90 per cent. Any small increment at this high percentage is valuable, but even a small increment is extremely difficult to obtain. Thus one can expect some improvements along this line, but nothing radical, barring of course one of those unforeseen inventions that cannot be taken into account in this prognosis.

COSTS OF BOOSTERS

The cost of boosting a payload to the high speed required for its mission —including the development of boosters, establishment of complex launching bases and operating them, and the manufacture of the hardware and the fuel—represents a major share of the total cost of the space program. In the first place, the development costs are huge; the boosters are complicated technological devices which require large design, development, and test teams to get them into any reasonable state of operational readiness. In the long run, however, development costs become less important than operational costs.

One of the major operational costs of space boosters is the fuel. Every vehicle, be it a long-range ballistic missile or an orbiting vehicle, takes off loaded as high as 90 per cent of its total weight with fuel which is burned in the mission. When one considers that these vehicles range up to 200,000 pounds now, and will range to millions of pounds in the future, one realizes that the fuel cost alone will be considerable. Rocket booster engineers know this full well; and in their search for high-performance fuel combustion for their liquid and solid propellant rockets they also keep an eye on the production cost figures of the propellants.

However, one should not be too discouraged by the fact that such a large percentage of the take-off weight is fuel. We already have experience with operations in which very large amounts of fuel are used but which have become economically feasible—for example, one of the standard jet aircraft used today by commercial airlines. With a take-off weight of about 280,000 pounds, the fuel weight of such a plane is 122,000 pounds, or between 40 and 45 per cent of the take-off weight. For a payload of the order of 36,000 pounds between a third and a quarter of the fuel weight is expended in a flight. If one considers not the typical passenger jet airliner but the long-range bombers designed for a maximum fuel capacity in order to achieve a maximum range, one finds that instead of between 40 and 45 per cent of fuel in the take-off weight, it runs to 50 to 60 per cent. So a mission which involves expending most of the initial weight of the vehicle in fuel consumption is not necessarily something that cannot be made economically feasible and even profitable.

In current space operations one of the greatest expenses arises from the fact that the booster vehicle is used for only one flight. R. C. Truax, Director of Advanced Development at Aerojet General's Liquid Rocket Plant, told a panel of space writers in New York that "if an airliner today were to be used only once on a cross-country trip and then thrown away, the fare per passenger just to pay for the airplane would be around $30,000." Clearly the practice of using a rocket booster for a space mission only once must be changed, an objective on which efforts are now under way.

Before describing some of the techniques of recovering the booster vehicles for space missions, it is of interest to establish in the reader's mind a cost figure for space operations to be used as a standard. It is a difficult figure to calculate accurately because a typical space mission involves not only the cost of the hardware and the fuel as purchased from the manufacturers but also the cost of the launching team and the team that tracks the vehicle in its flight, and so on. Since the organizations which carry out these functions are complex, it is somewhat difficult for a cost analyst to track through government organization and make a fair assessment to each of the many organizations involved for the cost of their share of the operation. Even recognizing this difficulty, engineers today use as a standard the cost of putting a single pound of payload into a circular orbit about the Earth at an altitude of about 300 miles. Rough cost estimates for various booster systems and various kinds of projects show that the costs run from $1,000 to several thousand dollars per pound of payload in orbit.

The long-range objective of booster engineers and designers is to cut this cost per payload pound in orbit by a factor of at least one-tenth and possibly by one-hundredth. It is not easy to achieve, at least by using the techniques at hand, but there are some hopes. A reduction in cost by the order of one-tenth might almost be achieved by making completely recoverable booster systems; there are many such proposals now under consideration. One generic type employs as the first-stage booster a kind of flying vehicle which, after taking off vertically and boosting the upper stages to some reasonable velocity, possibly a few thousand feet-per-second up to 10,000 or 12,000 ft/sec, converts itself into a flying vehicle which is flown manned or unmanned and landed as a conventional high-speed aircraft. Some designers would prefer to see the engines of this first-stage booster of the same type as current high-speed airplanes—namely, turbojet engines —reasoning that such engines give added convenience, reliability, and low fuel consumption. There are other proposals to use recoverable schemes involving parachutes and recovery systems such as snatching the returning launching vehicle in the air by a large helicopter. Though at first these sound complex and unreliable, more detailed examination indicates that they have some reasonable degree of feasibility.

Whatever the recovery system that is developed, one can be sure of two things: that such a system is entirely feasible, but that the actual development costs will be very large. It is the kind of complex development that requires few new basic scientific principles—only the application of a large amount of engineering design and effort. In the mind of the author, the development of such a system of the recovery of the early first stages in boosters is inevitable.

The cost per launch varies widely since different missions have different payloads. A few of the scientific missions of the past have used very small vehicles. For example, our first Explorer weighed only about 30 pounds, and our first Vanguard only 3 or 4 pounds. The first Mercury manned orbital capsule will weigh 2,000 to 3,000 pounds. Eventually tens of thousands of pounds of space vehicles will be sent into orbit.

For the small number of launch types of scientific and exploration missions the simple expendable booster systems will pay off best because their development costs are much less than those of recoverable systems. The many-sectioned solid propellant boosters will be somewhat lower in cost per launch than the liquid propellant systems, though as the number of launches increases, the liquid propellant systems approach the solids in cost. Also, as the number of launches go up, the cost of expendable systems goes down, but rather slowly. For missions where the number of launches begin to grow it soon becomes desirable to consider recoverable systems as discussed above—either those fully or even partially recoverable. The development cost for fully recoverable systems will be large, but not substantially more than for partially recoverable systems. With many launches, such as one might expect over the decades for a commercial satellite system or a manned military orbiting system, the fully recoverable systems would probably justify their initial development cost. In such high-performance recoverable systems which for a large number of launches will give the lowest cost per launch, one finds, as mentioned above, high-speed air breathing engines using the turbojets and ramjets and high performance liquid rockets; one also notes the entrance of the nuclear rocket into the discussion.

It is important that everything be done that is possible to reduce the high costs of space programs. The United States government alone is spending an amount approaching 10 billion dollars on its space research, development, testing, and operations programs—programs which of course are broadly based. The United States program as a whole includes medium- and long-range ballistic missiles, military and commercial satellites, scientific, research, and exploration missions penetrating deeply into the solar system, and scientific measurements of the characteristics of the Earth and the space around the Earth.

New boosters for space missions will cost hundreds of millions of dollars before they can be considered operational vehicles. After they are developed, the operative cost will still remain large until the devices themselves are made recoverable. But the present situation on development and operation is not radically different from that which existed in the development stage of large commercial jet transports. Those also cost hundreds of millions of dollars to develop, but large-scale operational use of jet transports has shown that the effects of the initial cost upon direct operating cost is almost negligible. It follows that it is important in space operation to be able to spread initial development costs over many operations if space is to become an important component of man's everyday life.

TESTING: THE BOOSTER EXPERIENCE

As developments in the component fields go forward in a technology as complicated as space flight, there is no substitute for actual space testing, the value of which has been demonstrated by the statistics which come out of the reliability studies on large space boosters.

Thus far in the development of space boosters the percentage of successes in the first ten firings ranges between two out of ten and six out of ten, with the average about four out of ten. From this rough average of forty per cent reliability for the first ten shots, reliability goes up into the eighty and ninety per cent region as the number of shots increases to a hundred. From there on, increased reliability in boosting gets more and more difficult and seems to be obtainable only with greatly increased numbers of firings.

Another indicator of the improvement that arises from experience is the *Box Score of United States Spacecraft Launches*. A similar box score for Russian launchings over the years is given. Here it is noted that the score is 100 per cent for all five years of the space age. This may be explained on the basis of somewhat different rules of scoring.

To back up the engineering developments listed above which are required to improve our space capability, some discoveries in the fundamental sciences of physics, chemistry, mathematics, and biology will be of help. In fact, in the forefront of research in the engineering fields the boundaries between these basic sciences and the engineering fields are fuzzy, and in the modern technological world they tend almost to disappear. This is not true, however, in the design, development, testing, and using of the large, complex space vehicles. Such activities are purely engineering. Here the engineer with purposefulness, vision, and sound technical training will lead the way.

The Nuclear Rocket

As one of the most interesting and widely discussed space developments, the proposed manned Moon exploration is usually publicized in the context of cost. The author has seen estimates of successfully landing a man on the Moon and returning him to Earth ranging from about one billion to one hundred billion dollars, with time estimates from 3 to 20 years. More responsible estimates range from fifteen to forty billion dollars, and from

BOX SCORE OF UNITED STATES SPACECRAFT LAUNCHES

1957

Vanguard TV3

Failures: 1   Successes: 0

1957 TOTAL: 1

1958

Failures	Successes
Explorer II	Explorer I
Explorer V	Explorer III
Vanguard TV3 (sic)	Explorer IV
Vanguard TV5	Vanguard I
Vanguard SLV1	Project Score
Vanguard SLV2	
Vanguard SLV3	
Project Able I	
Pioneer I	
Pioneer II	
Pioneer III	
Beacon I	

Failures: 12   Successes*: 5

1958 TOTAL: 17

1959

Failures	Successes
Vanguard SLV5	Vanguard II
Vanguard SLV6	Vanguard III
Discoverer III	Discoverer I
Discoverer IV	Discoverer II
Explorer Project	Discoverer V
Beacon II	Discoverer VI
Transit 1A	Discoverer VII
Atlas Able IV	Discoverer VIII
	Pioneer IV
	Explorer VI
	Explorer VII

Failures: 8   Successes*: 11

1959 TOTAL: 19

1960

Failures	Successes
Samos I	Echo I
Midas I	Pioneer V
Courier IA	Tiros I
Discoverer IX	Tiros II
Discoverer X	Courier IB
Discoverer XII	Discoverer XI
Discoverer XVI	Discoverer XIII
Project Echo	Discoverer XIV
Explorer Radiation Satellite	Discoverer XV
Atlas Able 5-A	Discoverer XVII
Atlas Able 5-B	Discoverer XVIII
Transit IIIA/GREB II	Discoverer XIX
Scout-3	Midas II
	Transit IB/GREB I
	Transit IIA/GREB II
	Explorer VIII

Failures: 13   Successes*: 16

1960 TOTAL: 29

1961

Failures	Successes	
Explorer S-45 I	Samos II	Discoverer XXIX
Discoverer XXII	Explorer IX	Discoverer XXX
Mercury Atlas III	Discoverer XX	Mercury-Atlas IV
Explorer S-45 II	Discoverer XXI	Discoverer XXXI
Discoverer XXIV	Transit III B/Lofti	Discoverer XXXII
Explorer S-55	Explorer X	Midas IV
Discoverer XXVII	Discoverer XXIII	
Discoverer XXVIII	Explorer XI	
Samos III	Discoverer XXV	
Discoverer XXXIII	Transit IV A/Injun/GREB III	
	Discoverer XXVI	
	Tiros III	
	Midas III	
	Explorer XII	
	Ranger I	
	Explorer XIII	

Failures: 10   Successes: 22

1961 TOTAL: 32 (to October 24, 1961)

* Payload successfully injected into orbit.

BOX SCORE OF USSR SPACECRAFT LAUNCHES

1957	1958	1959
Sputnik I Sputnik II Successes: 2	Sputnik III Successes: 1	Lunik I (Mechta) Lunik II Lunik III Successes: 3
Failures: ? **1957 TOTAL: 2**	Failures: ? **1958 TOTAL: 1**	Failures: ? **1959 TOTAL: 3**

1960	1961
Sputnik IV Sputnik V Sputnik VI Successes: 3	Sputnik VII Venus Probe/Sputnik VIII Sputnik IX Sputnik X Vostok I Vostok II Successes: 6
Failures: ? **1960 TOTAL: 3**	Failures: ? **1961 TOTAL: 6 (to October 24, 1961)**

From: "Space Log," Space Technology Laboratories, Inc.

seven to ten years. Granting the importance of cost, in the technical perspective the Moon program is of special interest because it points up the importance of the development of the safe nuclear rocket. Most of the Moon planning programs have been based upon the use of high-performance liquid-propellant rockets; and the cost and time estimates of all the development and the testing and the actual missions themselves have been based upon such rockets. But the liquid-propellant rocket has been selected only because its principal competitor, the nuclear rocket, is not considered to be in a sufficiently advanced state. If the Moon program objective were moved back to a 15- or 20-year objective instead of something less than 10 years, then clearly the nuclear rocket would compete.

Just where, then, does the nuclear rocket stand with respect to its promise for the future and its current development? There has been considerable publicity given to the nuclear rocket development; and in fact a joint National Aeronautics and Space Agency-Atomic Energy Commission development project with industry has already been started. This action follows a long period of experimentation on a test-bed nuclear rocket by the Atomic Energy Commission, aided by certain industries. The new developments are aimed at a flight test engine within a period of 5 to 7 years.

Just how important is the nuclear engine? One answer to this question is suggested by a comparison of liquid, solid, and nuclear propulsion systems in terms of specific impulse, which, measured in seconds, is a merit factor for the efficiency of the use of the propellant.[1] Liquid propulsion systems now are considerably better than solid propulsion systems, and they offer room for further improvement; but the specific impulse promised by nuclear rockets is far above anything that is promised by the liquid or solid rocket propellant systems. It appears that, while the best performance for a liquid propulsion system might be about 500 seconds, and the best for a solid propulsion system about 325 seconds, the best performance of a nuclear propulsion system might be a specific impulse of 1,200 seconds.

Actually, increases in specific impulse multiply over and over in the final performance of specific booster systems. That can best be shown from a graph (*Round Trip Lunar Flight*, etc.). In this graph the ratio of the gross weight at take-off to the weight of the payload is plotted against the number of stages for different kinds of rockets. If one designs a lunar rocket system requiring an impulsive velocity of 60,000 ft/sec, which is quite reasonable, and bases it upon the present liquid rocket specific im-

---

[1] Based on a graph, "Propellant Performance," taken from the testimony of Mr. S. K. Hoffman, Vice President of North America Aviation, Inc. and President of their Rocketdyne Division, and a leading developer of rocket engines, to the Committee on Science and Astronautics of the United States House of Representatives.

ROUND TRIP LUNAR FLIGHT WITH ATMOSPHERIC BRAKING

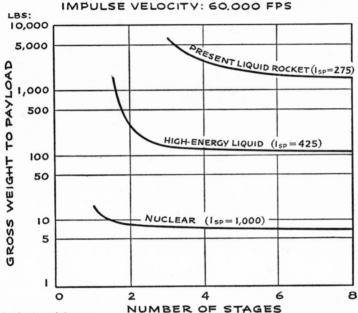

IMPULSE VELOCITY: 60,000 FPS

From: Douglas Aircraft Company.

pulse of 275, a six-stage vehicle would require about 2,000 pounds gross weight at take-off for every pound of payload. If the payload for a manned landing system were 16,000 pounds, the take-off gross weight would be about 32,000,000 pounds. By the time the lunar vehicle was developed, one could count on using high-energy liquid propellant systems with specific impulses possibly as high as 425. The graph shows that gross weight would be something over 100 times the weight of payload. With 150 pounds of payload the number of stages could be reduced possibly to three or four. The same graph shows that the use of a nuclear rocket with a specific impulse of 1,000 seconds would permit the use of about 8 pounds of gross weight per payload pound with only two stages.

The reader can get a good idea of the size scale of nuclear and liquid-propellant rocket space vehicles by the comparison shown in the accompanying figure (*Size Comparison*).

Clearly the nuclear rocket offers tremendous promise as a booster system for deep space operations. It is the author's estimate that the development of a satisfactory nuclear rocket booster system is the *sine qua non* for more distant future space operations. It is also interesting to note that the nuclear rocket does not require new scientific principles. The development

SIZE COMPARISON

From: Douglas Aircraft Company.

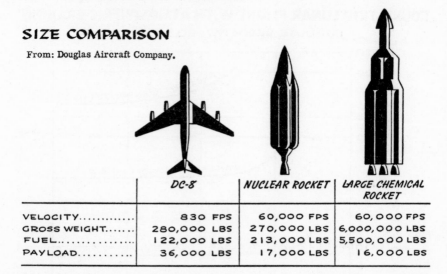

	DC-8	NUCLEAR ROCKET	LARGE CHEMICAL ROCKET
VELOCITY..............	830 FPS	60,000 FPS	60,000 FPS
GROSS WEIGHT.......	280,000 LBS	270,000 LBS	6,000,000 LBS
FUEL.................	122,000 LBS	213,000 LBS	5,500,000 LBS
PAYLOAD...........	36,000 LBS	17,000 LBS	16,000 LBS

of the nuclear rocket will be a long, complicated, expensive engineering project. If, as will almost surely be done someday, the nuclear rocket stage is developed in a form which is recoverable, it will open the door to a reasonable capability for operating in the solar system on rather elaborate missions to distant planets and returning without incurring overwhelmingly exorbitant costs. Such developments will take a long time. The author would not attempt to put a date on the achievement of these final potentialities except to say that it is clearly more than a decade away.

Some Major Problems

CONTROL, GUIDANCE, AND COMMUNICATIONS

One whole new field of development consists in the guidance and control equipment for space operations. Ballistic missiles can be guided by two basically different systems: the one employing so-called inertial navigation in which pre-set devices involving gyroscopes and accelerometers and computers guide the vehicle entirely throughout its flight; the other involving radio direction finding and radar distance measuring to perform continuous tracking and guidance of the ballistic missile. For very long-range space missions inertial guidance can be used in part, but there must be corrections to it made by optical or radio sighting devices to reference points such as the Earth, the Sun, and the other planets and stars. A wide range of devices which contribute to this lore has been worked on over the recent decades, and the problem seems to be one not of discovering new principles but of making the technical advances necessary to develop new equipment.

RELIABILITY

In the types of devices needed for communications, guidance, and control there are many electronic, mechanical, and thermal parts which add up to very complex systems. Furthermore, there are restrictions concerning weight and size and power consumption of these parts. Experience with such complicated systems has shown that long-term reliability is always a problem.

Long-term reliability can be obtained for very simple devices. For example, the much maligned Vanguard program, which was hopefully our first but turned out to be later in the series of satellites, and the first of which consisted of only a three-and-a-quarter pound, 6-inch diameter sphere with a shell of aluminum containing two very simple radio transmitters, has been operating in space since March 17, 1958. Since the orbit it attained is sufficiently high so that the drag of the Earth's atmosphere does not tend to slow it down, the vehicle is expected to orbit the Earth for a long time, possibly centuries. One of its two transmitters was still broadcasting its position after 3 years—but it must be understood that this radio transmitter is the simplest of devices. The problem lies in the fact that reliability tends downward rapidly with increased complexity, and that thus far the process of making reliable the complex guidance, control, communication and other complicated equipment for spacecraft is difficult.

Some indications of the problems of reliability of communications equipment are shown by the tracking of the Sun satellites which have been established by both the Soviet Union and the United States. For example, the Pioneer V had the record interplanetary distance radio communication of something over 22,000,000 miles, before its communications equipment failed. The late Soviet Venus probe failed to transmit after going only a fraction of that distance.

There is one school of thought which believes that the best way of handling the complicated devices for space flight to ensure reliability is to have trained men aboard the spacecraft to repair them. This argument is advanced as one of the most important reasons for putting men into space. In the author's estimate, reliability of space equipment cannot be attained by repair and maintenance operations. One of the biggest problems yet to be conquered in space, it can be solved only by the slow process of improvement of design.

THE SPACE ENVIRONMENT

Whenever man has contemplated going into a new environment—sailing far from his native shores, first going up in balloons and airplanes—he has been challenged by the difficulties, real and imagined, of the new world he is entering. And so with space. For the space environment differs from our environment here on the surface of the Earth. There is no atmosphere to shield humans and equipment from the physical bodies, space particles and electromagnetic radiations in space; there is no atmosphere to supply life-giving oxygen.

Some of the harmful radiations such as the ultraviolet radiation from the Sun which would burn skin and eyes seriously can be easily shielded with only a small amount of material, such as the structural skin which any spacecraft requires. There are, however, other radiations from the Sun which are much more dangerous and occur mainly in solar flares. For example, there is an extremely high-energy flare from the Sun that might occur, say, once every four years, in which the energy of the particles range in the 370,000,000 electron volt range which is extremely dangerous to humans and from which they would definitely have to be shielded unless one wanted to take the chance that no human would be exposed when such a flare occurred. There are other solar flares which occur once a month or so and give out heavy radiation, concentrated around 46,000,000 electron volts. These are not as intense but still have to be taken into account. Around the Earth there is a belt of charged particles called the Van Allen Belt. The energies of the particles concentrate around 144,000,-000 electron volts and also are of sufficient number that they must be taken into account in shielding. From outside the solar system, from galactic sources, there are cosmic rays with extremely high individual particle energies—around 4,000,000,000 electron volts—but which come in smaller numbers than the others. Finally, if the spacecraft employs a nuclear rocket, there must obviously be shielding for the direct and scattered neutrons and direct and scattered gamma rays from the nuclear reactor.

Throughout the solar system there are very fine particles of dust called micrometeorites, and, scattered in much smaller number, particles larger than dust. The distribution of these particles indicates that there will be some problem due to the slow weathering of outer surfaces of space vehicles by the impingement of this dust, which has a sand-blasting effect. Collision of a spacecraft with larger particles would create a hole, but self-sealing techniques and design of multiple-layer skins can minimize this hazard. Possibly the best way to describe the engineering problems raised by the foreign environment of space would be to list the various weights required in a typical vehicle design for a three-man spacecraft intended to travel extensively throughout the solar system and having a total weight of 52,000 pounds. (This vehicle has not yet been developed, but reasonably complete engineering studies have been made of the system.) The following figures, taken from Douglas Aircraft Company reports, are in pounds:

Pressurization and Oxygen System, 630
Thermal Conditioning, 720
Atmospheric Control, 340
Space Suits, 270
Three Men, 600
Interior Equipment, 560
Earth Survival Pack, 234
Food and Water, 348
Structure, 2,000
Shielding, 10,000
Electronic Equipment, 1,022
Power Supplies, 1,320
Last Stage, 20,000
Cargo, 14,000

One can see from this summary that the items required to provide the proper thermal and atmospheric control for men are relatively small; even the food and water become small items. The big items are shielding, any cargo or equipment necessary for the men to take on a trip through the solar system, and last-stage propulsion devices. But such a total can be handled by the very large liquid propellant rocket systems and the nuclear rocket systems under development.

Prospects

All that has been said here is quite independent of the special features of national technology, or of the Soviet-United States competition in space. It is clear that Soviet technology has been able to accomplish space missions with boosters larger than those used so far by the United States. However, long-term progress for the Soviets no less than the West will depend on the same considerations spelled out in this chapter.

It is this author's conclusion that, although the problems are many, the currently contemplated space missions are technically possible, and even the hazardous new environment of outer space presents no conditions which are impossible to counter by modern technology. True, the development of all the equipment for providing safe flight for humans in space will be an expensive development program, but on the other hand it seems to be a reasonably straightforward one.

In the end, achievement of the capability to use space profitably for mankind will depend upon the slow, expensive accumulation of engineering experience, not on spectacular break-throughs in the realm of scientific principles.

2.

Prospects for Human Welfare:
Peaceful uses

Introduction

♦ DONALD N. MICHAEL

Ever since Sputnik I there has been a plethora of free and easy predictions about the social impact of space activities and about the various ways in which space will substitute for, replace, or extend man's present earth activities. It seems worthwhile therefore to begin this discussion with some comments about what kinds of predictions will and will not be attempted herein and how these will differ in spirit and range from those which have been so easily made by so many for so many different reasons.

In the first place, we shall look at most no more than about 20 years ahead. But even 20 years may be too far to stretch the imagination if our speculations are to be more than sheer fantasy. For the pace of social, political, and

31

♦ DONALD N. MICHAEL, a social psychologist with a background in the physical sciences, is professionally interested in the relationships between technological change and social change. Before *Sputnik I,* he began drawing the attention of his colleagues to the social implications of astronautics. Recently, he directed and was chief author of the Brookings Institution publication and report to NASA, *Proposed Studies on the Implications of Peaceful Space Activities for Human Affairs.* He is now Director, Planning and Programs, of the Peace Research Institute.

especially technological change in areas other than space is such that man's goals and his means for attaining them will also change radically even during this short time. We need only mention as examples the population explosion, the widespread use of computers and automation, and the growing aspirations and power of the many new and underdeveloped nations.

Since we are creatures of present perceptions, present values, and present environment, we can do relatively little to explore and speculate how a later world will feel and act. The further ahead we look, the greater the chance that we shall, except by good luck, miss the target widely. Hence we limit our speculations to a time period in which our present perspectives, values, and environment are still meaningful.

Second, there are so many conflicting and unresolved criteria for the allocation of resources for space activities that most forecasts about time and cost are meaningless. They are almost always optimistic, in part because they are made mostly by people whose reasons for going on record vary from naive exuberance to special pleading for social and financial support, in part because they overlook the social costs of introducing new space developments—the displacement of existing capital equipment, personnel, and sometimes personal power. They overlook as well the costs of developing the complementary technologies and the costs of training and buying the necessary manpower. After all, there is no reason why, given the present technology, we should not have a totally automated post office system; nor is there any technological reason why all railroads and airplanes should not run on time. That this is not so is due to a complex of forces and circumstances, more or less evident to the reader; analogous conditions will in analogous ways affect space activities.

Hence the reader will find in this chapter no estimates of the cost of a letter by rocket to the moon or of a first-class seat on a rocket from Chicago to Moscow or even the cost of a telephone call via communications satellite from New York to New Delhi. None of the estimates for such space activities ever adequately examine development and installation costs and all the other costs of maintenance and operation that need to be met to fit the technology into an ongoing society.

Third, technological developments already under way, along with the social forces which affect them, need especially to be kept in mind. For example, by the time the long-range weather forecasts are made possible, in part by weather satellites, many now weather-dependent human activities may be no longer dependent on weather. Thus weather forecasts may not have all the exciting implications that we shall discuss presently. Similarly, some anticipated uses of communications satellites may be different from those suggested here because of the development of other high capacity, inexpensive communications systems, some of which are now in the research and exploration stage. So I shall from time to time refer to alternative technologies which may radically affect the potential contributions of space activities to society "down here"; and the reader will do well to be alert to opportunities for further qualifying our predictions and speculations. Of special importance in this regard, full appreciation of the peaceful use of space depends on appreciating the other aspects of space activities examined elsewhere in this book.

Implicit in the approach of our discussion are two other attitudes which merit brief statement. In the first place, the consequences for society of peaceful space activities are seldom treated here as simple cause-effect relationships in the fashion, for example, of the optimistic equation that space research equals better weather forecasts equals progress equals greater happiness. Such simple-minded statements, while popular and prevalent, are essentially useless at best and misleading at worst. Whether or not space brings greater happiness or convenience or anything else depends not nearly so much on what happens in space as it does on what happens when space activities impinge on human society. Thus most of our attention, once we have indicated how space provides the potentiality for new uses of knowledge or processes, will be on the non-space aspects of the problem. With this approach, space activities as such tend to disappear into the background while we examine the ways men may wrestle with the problems always produced when new knowledge confronts old traditions.

This leads to the second point: many of the problems and opportunities which space will provide are not new. Rather, they are old ones made more pressing. There is a great tendency to hope that the conquest of space will solve a host of present problems. The cold hard evidence demonstrates that instead it will more implacably confront us with them.

Communications

PRINCIPAL TYPES OF SYSTEMS

Appreciating the implications of communications satellites requires first an appreciation of some of the major characteristics, and the assets and liabilities associated with these characteristics, of the two chief types of communications satellite systems. These characteristics have profound implications both for the economics, politics, and utility of satellite communications and for our understanding of the social problems and opportunities involved in their use.

The "passive reflector" satellite, such as *Echo,* typifies one kind of communication system. This satellite contains no electronics and simply reflects or bounces signals from a ground-based transmitter to a ground-based receiver. To do so effectively, however, requires very high power transmitters—perhaps two to ten times as powerful as those typically used in commercial broadcasts. It also requires highly sensitive receiving antennas, around 250 feet in diameter, which must at the same time track the moving satellite rapidly and accurately. The home viewer or listener can receive messages relayed by this system only through a central receiving and local retransmitting installation, not directly from the satellite. On the other hand, the passive satellite has the great advantage of unlimited two-way channel capacity over a very wide band of wavelengths. Also, if improvements in communications technology should make it desirable to introduce changes in ground equipment or in the characteristics of the transmitted signal, this could be done without changing the satellite.

Anticipated technology would permit a transmitting and receiving area with a radius of about 4000 miles. Thus global coverage would require at least 12 satellites orbiting around the world; these would be subject to air drag, solar deflection, micrometeors, etc., all of which would be deleterious to satellite life and satellite orbit characteristics. Finally, since these satellites are passive reflectors, and anybody with the appropriate ground-based equipment could use them for transmitting and receiving, control of satellite use and frequency assignment could be hard to enforce. The possibilities are good for accidentally or deliberately jamming the received signals from reflecting satellites by simultaneously transmitting more than one signal into the same receiving area on the same frequency.

The other type of communications system uses an "active repeater" satellite, which contains its own receiver and transmitter and the power supply to operate them. Consequently, the ground transmitter can be as small as perhaps 100 watts, and private parties can in principle receive the re-transmitted signal from the satellite with nothing larger than a fringe area antenna. However, if a user nation is willing to give up the advantage of direct reception from these satellites and use large and costly antennas similar to those for passive satellites, the capacity of these active satellites can be increased greatly and frequency interference reduced as well. Under these latter conditions the satellite can broadcast to the big antenna a much weaker signal than can be directly received. This weaker signal requires less power and less massive transmitting equipment: the weight saved can be applied to providing more channels. And less power also results in less interference. If the satellites are placed in appropriate orbits at 22,300 miles altitude—the so-called hovering satellite—three of them can cover all but small parts of the polar regions in a world-wide umbrella of communications.

These active satellites, in contrast with the passive, will have a limited capacity for receiving and transmitting messages simultaneously, and the messages will have to be transmitted at the precise frequencies built into the satellite. Thus, scheduling message priorities and wave-lengths will require complete collaboration between users as well as centralized control and compatible equipment. Also, changes in the frequencies at which the messages are transmitted, beyond those frequencies built into the satellite, or changes in the characteristics of the message-carrying waves themselves, would require satellite replacement. Obviously these satellites, weighing anywhere from hundreds to thousands of pounds, will be much more expensive than the passive satellite, which is simply an inflated sphere of metallized plastic. And their active life must be longer than that of the passive satellite to make them economically feasible. Moreover, the research and development to make them reliable long-lived systems will be costly and time-consuming.

IMPLICATIONS OF COMMUNICATIONS SATELLITES

Telephony—With present channels hardly adequate to handle today's message load, current projections show vastly increased demands for trans-oceanic telephone circuits over the next two decades. If the demand is to be met, advanced communications systems will be necessary. At the moment, so far as over-all development and operating costs have been calculated—which has not been very far—it appears that communications satellites would be a cheap answer to this demand.[1] (However, in many of the laboratories working on communications satellites there are also contending communications systems being developed, including advanced means for carrying many more messages per cable, transmission by reflection from the troposphere, and even the reflection of radio signals off the burning meteorites which endlessly and multitudinously bombard the earth's atmosphere. Sophisticated methods for reconstructing messages may also radically reduce the channel capacity need per message and thereby greatly increase the capacity of the available system.) Whether or not the actual number of channels will be great and cost per message low—as has been frequently claimed—will be determined by the ownership and operating philosophies implemented. But if rates are low and capacity high, the opportunity for frequent, rapid voice communications should open new possibilities for increased control and coordination.

---

[1] See Chapter 3.

Among other things, increased communication capacity may give bigger organizations, whether governmental or private, a further capability to absorb small organizations. On the other hand, easy communication may increase the ability of small organizations to compete. We do not yet know enough about the relation of communication to organization to know which will happen. In any case, it is likely that the world will be increasingly dependent on the maintenance of a far-flung communication net. The consequences of disruption of such a net are hard to visualize. Whether they would be any worse than a world-wide equivalent of New York buried under 20 inches of snow remains to be seen.

One other kind of international communication might be especially affected by the ready accessibility of telephone circuits. The number and activities of international organizations certainly will not decrease in the years ahead; and their growth, stability, and effectiveness may be increased by the opportunity for opposite numbers in various nations to keep in touch by phone when working out problems. Whether or not such easy access would be to the advantage or disadvantage of what some believe to be an already too rapidly communicating diplomatic community, it certainly would be to the advantage of those organizations and those parts of the diplomatic apparatus not concerned with day-to-day crises. Such informal communications could, over time, lead to an increasing acceptance of international goals. Here, then, is one area where international communications are likely to induce a world viewpoint in place of parochial national viewpoints.

Data Search, Retrieval, and Processing—The trend of transmitting large amounts of data to and from computers and users via telephone should gain impetus on a grand scale by using either satellites or the ground-based communications facilities freed by the shift to satellites. In cheap, rapid, world-wide access to computers and their memory banks there are vast implications for both good and evil. Far-flung systems could be controlled from centralized computer facilities; widely distributed memory systems and computers could be brought to bear locally on special problems. In principle at least, the resources of all the world's libraries, their works so coded as to be automatically searched on command, could provide distant scholars with automatically transmitted data otherwise difficult to obtain. Elaborate monitoring systems for operating arms-control programs through high-speed correlation of information from widely dispersed sources would be another use.

But perhaps the two greatest consequences of a world-wide data processing potentiality are more in the realm of states of mind than in that of particular applications. In the first place, as frequently happens in this world of incessant technological advance, invention becomes the mother of necessity. The very fact of potentiality for data processing on a grand scale will encourage the invention of needs for such data processing, needs which might even now be met with present communication systems but are not now recognized or acknowledged, in part because they are not so evident as they will be when seen in the light of the communication satellites. One can imagine the proliferation of practically instantaneous reports on everything from the world distribution of diseases to inventories of fuel; from the availability of various types of professional personnel and workers to up-to-the-minute analysis of world-wide attitudes on the actions of governments. And the very fact that such information is at hand will almost necessarily lead to means for using it; decision making and planning will be both speeded up and complicated. What this implies for the decision maker in government and industry is not clear. But apparently it will demand a different type of executive, sensitive to different types of information and working in a different time framework than is now typical.

In the second place, as with our own brains, we now use only a tiny per cent of the potentialities inherent in our external data processing systems. Much of what we learn dies on the shelves where it is filed away, undiscovered and unappreciated by those who could use it. The processing and data-using capacity provided by satellites may stimulate a long overdue attack on the development of methods for using the data we have.

Three related implications of this potential revolution deserve mention. If data processing, numbers, figures, and patterns are communicated either by wires freed by communications satellites or by the satellites themselves, there will be a substantial reduction in the messages communicated on paper, with consequences possibly disastrous for those organizations which exist to transport information on paper. Moreover, all such data processing potential involves the development of other technologies and specialities. Needs for computers, for human coders and programmers, and for maintenance and repair personnel would increase tremendously, as would the need for training facilities.

It is possible, of course, that demands would be so great that television channels, which require immensely broader band widths than do data processing and telephony, would be hard put to obtain a share of the satellite frequency spectrum adequate for their own uses. This might be especially so if the communication system is operated for private profit or as an important source of government income. Deciding whether one television channel for teaching purposes is worth a large number of telephone circuits will not be easy in a world of conflicting values.

With the enmeshment in an ever expanding computerized society will come an intensification of the social, ethical, and moral problems and opportunities inherent in communication and control. Many scholars and observers of technologyzed society have commented on these problems, so we shall not reexplore them here. But we would note that there will also be other, possibly profound, problems associated with a breakdown or disruption in the communications system spawned by the availability of communication satellites.

Mail—There may be a demand for high-speed mail transportation using facsimiles transmitted by satellites. The components already exist for scanning the letter, sending it electrically to a receiver, and printing a facsimile for delivery. The satellite would extend the length of the transmission link and, in principle, could transmit a tremendous number of letters rapidly. If this should become a popular form of communication, present letter carrier vehicles would lose a steady source of income.

Entertainment—It is now generally believed that entertainment by world-wide television will be economically competitive with the potential telephony and data processing. Fantasies abound about sitting by the three-dimensional, color television set for a live presentation of the Bolshoi Ballet from Moscow or vaudeville from London or a spectacular night view of an eruption of Mauna Loa from Hawaii. These fantasies need to be matched with some hard facts of reality. The first is the differences in the audio-visual imagery of nations and cultures, not only in language but also in format, pacing, style and taste. Such differences limit the acceptability in some countries of programs prepared in others.

There are complex economic and political questions of how international programs and transmissions are to be paid for. Not all countries permit costs to be defrayed by advertising revenues; and where they do, advertising which is permissible and economically desirable in one country might compete with the interest of another country in enhancing the home markets for its own products.

Even more sobering is a physical fact which no amount of compromise can change. When it is 8:00 p.m. in New York City, it is 4:00 a.m. in Moscow, 1:00 a.m. in London, and 3:00 p.m. in Honolulu. Given these time differences, many attractive and even profitable live programs will have to be video-taped after all. Perhaps it will be efficient to beam the live performance by satellite to ground-based studios for taping and local retransmission at appropriate times. It may be even more efficient to fly the video-tapes, made on site, by fast jet to already existing local transmitters.

Education, Information, and Propaganda—The question of using TV from satellites for large-scale teaching programs in underdeveloped areas is more complex than it is usually thought to be. The cost and character of receivers and the antennae needed for direct reception of satellite transmitted TV are central problems. A related problem is maintenance of the local receivers. But overriding these technical contingencies is a psychological problem—that of implanting incentives to learn from TV transmissions in people previously unexposed to the social and psychological imperatives associated with the systematic learning necessary for participation in modern societies. The problem of developing TV teaching materials for such people is equally formidable. We are only beginning to understand the implications of teaching by TV in our own society, to say nothing of applying our knowledge of them to other societies. Thus, other things being favorable, education in underdeveloped areas by satellite TV will be contingent on large-scale research programs on teaching from a TV screen. Perhaps space activities, then, will stimulate interest in more systematic research on means for effective communication between diverse cultures.

If satellite TV channels are to be operated for financial profit, then as long as high-paying demands for the channels exceeds the supply it is hard to imagine extensive investments for teaching in underdeveloped areas. On the other hand, ownership of at least some satellites by an international public service agency could solve this problem. There would also be an opportunity for the richer countries to help the poorer ones carry the costs of the satellites. If, then, it should turn out that important subjects can be taught by TV and taught more efficiently over-all than by other ground-based means, and if the economics and politics of ownership permit, satellites could have revolutionary social impact in the teacher-starved areas.

As for teaching in the more developed areas, there is the question of what can be better taught by satellite TV than by more conventional means such as film, video-tape libraries, and airborne TV. To be sure, the live presentation of world-shaking events would benefit students; but such events are likely to be infrequent and may be just as effective if presented locally via video-tape a day or so later.

Very large areas can be covered simultaneously by satellite TV, but this in turn could require scheduling of teaching periods according to time differences. Also, to some extent at least, extensive teaching via satellites requires a congruity of teaching philosophy and subject matter between regions which may prefer different philosophies and subject matter. However, satellites might provide an inexpensive means for schools to keep their teaching video-tapes up to date. During low-demand hours central libraries of video-tape could transmit new subject matter to local school receivers. And since video-tapes are erasable, schools would not need to maintain their own video-tape library.

As to general education of students and non-students alike, many seers predict the world-wide TV will increase mutual tolerance and understanding. Present evidence, however, indicates no certainty that exposure to other ideas necessarily increases tolerance: it may in fact increase indifference or hostility. Thus, while world-wide TV could enlarge perspectives, whether it will or not will depend more on what is presented and on what we learn about the psychology of changing attitudes and values than on the technological capability itself. Hopefully, the very existence of this capability will encourage exploration of these psychological factors.

It is, of course, only a short step from education to propaganda and political manipulation. Without adequate controls, intensive propaganda efforts are likely to be a major use for communication satellites and, as such, a major source of international friction. This is one of the strongest arguments for internationalizing at least the TV and radio components of satellite communications systems.[2]

Finally, it is worth speculating that communications satellites may free sufficient ground-based capacity to make closed-circuit conferences economically feasible when compared to the costs of transporting people to "in-the-flesh" conferences. If such conferences should become popular, the outlook for transportation, entertainment, and hotel industries may be serious. Perhaps the first faint implications of the changes might be found in the recent speculations that TV has outmoded at least the style and format of the national presidential nominating convention.

POLICY QUESTIONS TO BE RESOLVED

The technological characteristics of communications satellites have important consequences for their ultimate fate: they affect the ways in which the satellites can be used, and the ways in turn affect the economics and politics of satellite communication development and operation. Inherent in the broad geographic coverage and high channel capacity are the following problems:

1. There must be an international solution to deep differences over the proper allocation and sharing of radio and TV frequencies. At home there is already conflict between those who want full use of the optimum satellite frequencies solely for satellite communication and those already using the frequencies for a variety of ground-based activities such as remote control of mechanical systems and communication between vehicles. Internationally, the situation is approaching chaos, with the desirable frequency ranges saturated and overlapping, and the pressures increasing from those who want to share what is now available. The situation is by no means insoluble since the same frequencies can be used in many different places by local transmitters with their limited ranges; but communications satellites, because of their broad coverage, will require more careful assignment and control of frequencies than at present. As of today, allocation of frequencies is for the most part under the jurisdiction of the International Telecommunications Union. However, this Union has no jurisdiction over administration of radio frequencies; nor has it had to cope with the kind of allocation problems which will arise when many nations will be competing for the same frequencies at the same time while they disagree about how those frequencies should be distributed as between telephone, radio, television, and facsimile transmissions.

---

[2] See Chapter 7.

2. International agreements will be necessary to assure compatibility of equipment components used by various nations. These are presently produced by several manufacturers around the world and operated according to the standards of the nations involved. Here the experiences of the Euravision, which over the years has gradually extended a cooperative and integrated European TV and radio system, will be of great use. However, even between European nations the process has been difficult. This adds emphasis to the problems that will arise when many other nations are part of the system and there are competing systems. Frequencies for TV transmission from satellites differ from those used with conventional television receivers, and American and European receivers use different numbers of scanning lines per inch. Capabilities for direct reception of both conventional and satellite signals could be built into new receivers. But it is not clear that there would be enough interest in satellite-based reception to produce the large market which would make worthwhile investment in a system of direct reception by individual receivers. So far, color TV has been a financial disappointment to its proponents. Even so, its potential attractions seem to be greater than those to be offered in the early days of world-wide TV.

3. International problems relating to the assignment of privileges and priorities for satellite use will need solution. If the satellites are passive, then any organization with transmitters and receivers can use them regardless of the wishes of those who have paid for putting them in orbit and maintaining them. Also, passive satellites can be made to jam or override other signals received from them in the same areas at the same time. If the satellites are of the active repeater type, then very careful scheduling will be necessary in order that incoming messages do not overload receiving and transmitting capability. Just how it is to be decided, and who is to decide the criteria for the privilege of transmitting on a given frequency at a given time, have yet to be faced. Just as there are problems regarding privileges and priorities for satellite use, so there are analogous problems connected with transmitting and receiver equipment and antennae.

4. The purpose of satellites in the first place is to reach particular audiences with particular messages, whether these be entertainment, propaganda, information, or education on a local or worldwide scale. But there are deep differences among national philosophies as to the purposes of telecommunications as well as differences within nations (viz., our perennial arguments on the proper use of television). The resolution of these differences, and their reflection in regulations regarding the allocation of priorities, time, and substance to various patterns of programming, will be difficult and time-consuming.

For the United States two major questions subsume all of the problems posed above. Who in the United States will pay for our home-built satellite systems, the government or private industry? Who will control and who operate the systems? If, for example, it is decided that satellites can better serve mankind through international rather than private ownership and control, then the role of private enterprise in this country will be quite different. Still to be resolved, too, is the appropriate place for the private-profit incentive, compared to the larger public interest as represented by the government, in designing, building, and operating the systems. In many countries communications systems are government-owned or government-subsidized. As such they are subject to the philosophies of public service rather than so largely to the philosophy of private profit as in this country.

The Soviet Union very likely will develop its own system. (Eventually a group of European nations may contribute a smaller system as well.) Competition or cooperation between ours and theirs and the pressures that other nations bring to bear regarding the operation of the systems and their participation in them will have much to do with determining the most effective means of ownership and control.

Weather

There is some difference of opinion among atmospheric scientists as to when meteorological theory and practice will permit more precise short-range weather forecasts as well as world-wide, reliable, long-range predictions at least a season in advance. However, whether it will take a decade or longer to collect the data necessary for major theoretical developments, it is generally expected that in the years just ahead satellites will make major contributions to predicting such special phenomena as hurricanes, typhoons, and tornadoes, especially in the many parts of the world which now have no ground-based weather facilities for forecasting. In addition to weather satellites, rockets probing deeper into space will contribute solar data which may well be crucial for the development of weather theory, and communication satellites will probably be needed to handle the immense amount of weather data which will have to be processed at centralized computers and distributed to regional users. But it should be understood—and herein lie some of the special implications of space activities for weather forecasting—that weather satellites and rocket probes cannot by themselves provide the wherewithal for world-wide forecasting. They will be necessary parts of a system, but an elaborate ground-based network for data collecting, processing, and distributing will also be essential.

Before further discussing satellite-weather forecasts it should be emphasized that we shall be dealing with weather *forecasting*—not weather *control*. In the more spectacular descriptions of space activities weather control and weather prediction are frequently confused. It is generally agreed by meteorologists that, except for occasional and limited efforts at fog dispersal or local rain making, weather control will probably require, at the minimum, precisely the kind of knowledge about the atmosphere and the earth necessary for developing the theory needed to do the kind of forecasts we are going to speculate about. It appears possible that weather forecasts can be improved substantially; there is no clear evidence that even in principle is it possible to control weather on a large scale. Ecological warfare based on changing rainfall patterns and hurricane and typhoon dispersal are, if at all possible, probably much farther off than our twenty-year view of the future.

Better forecasts may give rise to new problems and opportunities as well as accentuate some old ones which will press even more for attention under the impact of new technology. But before looking at specific possibilities, two general points need to be kept in mind. In the first place, improvements in predictions will come only gradually. There will undoubtedly be a period when uncertainty about the reliability and accuracy of weather forecasts will also produce uncertainty as to whether it is worth investing in the economic and social costs necessary to gain the benefits if the forecast should be correct. Even in the United States, which is used to fairly reliable weather predictions, people vary greatly in the extent to which they consider such predictions in making important decisions. In countries where people are not used to systematic planning in terms of weather forecasts, and where the society and economy are not geared to responding to such information, the economic and social costs will be great. Therefore the introduction of a way of life which responds to better weather forecasting is likely to be slow, certainly until forecasts have demonstrated their accuracy sufficiently to make it worthwhile betting on them. Secondly, many activities are no longer critically weather-dependent, and it is possible that in the years ahead they will become even less so. Such developments as air conditioning, desalinization of sea water, and high precision transportation guidance may substantially reduce the impact of weather and thus lessen the social implications of weather forecasts based on satellites.

IMPLICATIONS FOR PRODUCT RAISERS

On the face of it, better weather forecasting, especially on a world-wide, long-range basis, should have profound effects on all primary food producers. As the world's population grows, demands for food will increase, and there will be strong incentives to raise edible products wherever and whenever weather permits. Population pressure may also inspire major developments in product raising or food manufacturing which would in some regions free them from weather dependency. Synthetic photosynthesis, sea farming, food from algae, food grown under plastic, hydroponics, irrigation techniques expanded by desalinized water distributed by nuclear-powered pumps—all could, in principle, essentially eliminate the crop-raisers' present dependency on the weather. But even if such techniques should become ready during the next twenty years, many areas of the world would still depend primarily on the weather for their crops. Hence if space activities provide the wherewithal for better weather forecasts, they will be useful at least in those parts of the product-raising world not now having the benefits of weather forecasts.

But what seems reasonable and worthwhile to one society does not necessarily seem so to another. Those areas of the world likely to gain the most from weather forecasts are also the most tradition-bound. Their people are unfamiliar with the meaning of scientific forecasts and scientific product raising. Among the less sophisticated, traditional crop growing rhythms may differ widely even among groups living in essentially similar climates. We can therefore expect varying responses to suggestions that the dates of planting foods be shifted for increased yield. Some may object to rotating or changing crops on the basis of unfamiliar types of forecasts about the amount of rain or the average temperature expected for the coming season. Moreover, changes in growing patterns mean changes in marketing and distribution patterns and in food preference standards. The whole economy and "style" of the society would be altered, especially if it is a one-crop economy, as is now often the case. With these thoughts in mind, let us look at some of the specific consequences for food raisers which will probably be associated with reliable seasonal forecasts.

Using such long-range forecasts would involve more than the willingness of the product raiser to plant, fish, or graze his animals according to the forecast. His actions would also depend on how easily he could get credit to buy seed, food or nets, or equipment for irrigation, smudging, sheltering, etc. This in turn would probably mean that governments would have to develop policies for advancing credit quickly, for educating the product raisers in new husbandry techniques, for providing storage schemes, and so on. In particular, it would face all governments with the problem of how to plan and allocate crop levels on an international basis in the light of forecasts of shortages in some areas, of multiple bumper crops of the same products in competing nations, and so on. It would appear, then, that to the extent crops are raised for export or are purchased by import, product raising would no longer be an activity to be decided chiefly on the basis of domestic considerations but would be based on international agreements and arrangements planned a season in advance.

In areas where they have not been used before, weather forecasts will require special personnel with expert knowledge of: the meanings of weather predictions, the relationships between products and other weather-related factors such as insects and local customs, and government resources for taking advantage of weather predictions.

Since there are no such experts now, and there are probably not enough meteorologists for the many sophisticated tasks which a world-wide weather system will impose, people from many countries will have to be trained. There will be problems and opportunities for increased international cooperation in establishing training institutes and meteorological stations. As yet, however, no adequate attack on these problems has even begun.

IMPLICATIONS FOR TOURISM AND RECREATION

Improved medium-range forecasts would most certainly affect the scheduling of spectator events such as mass meetings, sports, and other outdoor entertainment. The economic and legal aspects of changing in the light of later forecasts dates scheduled far in advance would complicate life for those who arrange such activities.

Long-range recreation planning is another matter. Week-end recreation is already profoundly affected by local, short-range forecasts; and in principle at least, long-range forecasts might have similar consequences. Whether or not they do will depend on what factors in addition to weather determine when people take their vacations. For example, if present leisure trends continue, along with the population growth, there may be a high demand for tourist facilities at all times. Since seasonal forecasts will for the most part refer to the general character of the weather rather than to whether or not a given day or week will be sunny, people may continue to put their emphasis on the availability of accommodations and hope for a break in the weather.

On the other hand, forecasts about the general character of the expected weather could be especially important for regions where recreation is a major industry, such as the Riviera, Florida, or the ski regions of Canada, Switzerland, and New England. Seasonal forecasts of exceptional weather could have disrupting economic, perhaps even international political consequences, if, let us say, the United States were to predict rain for the Riviera and sun for Florida during a given winter.

If and when long-range predictions do have an impact on the recreation industry, the effects may produce more instability in the industry than it is now subject to, with exceptional overloads in good weather and insufficient loads in bad. Thus there might be changes of ownership in the tourist industry, with various combines carefully located geographically so as to increase the likelihood of balancing losses with gains. One can also imagine attempts to attract tourists through guarantees to move them from one place to another if the weather turns out to be as bad as predicted. Already there are trends toward alternate recreation facilities not dependent on weather.

Certainly we can expect some strong protests. As the quick demise—

under pressure from businessmen—of the weather bureau's "discomfort index" indicates, better weather forecasting will not always be to everybody's benefit.

IMPLICATIONS FOR TRANSPORTATION

Over the next twenty years transportation will become even less dependent on weather variations than it is now, with navigation in all types of weather made safe by guidance devices originally invented for space exploration. Various modes of transportation would still be subject to destruction or immobilization under unusual weather circumstances; even now the extent of disruption of motor, rail, and air travel appears to be more a matter of resources to cope with the results of bad weather than of foreknowledge that bad weather is coming. However, if one knew a season in advance, or in some cases even weeks in advance, that the winter— or particular days—were going to be unusually bad, communities might be able to appropriate funds to keep transportation operating. Here, of course, the standard problem arises—whether avoiding loss of a few days of efficient transportation is worth the costs of having on hand the equipment to meet the exceptional few days. Solving this question involves many of those conflicts of interest which will determine how much is made of the opportunities afforded by space activities.

Freight transportation systems will find it advantageous to have precise advance estimates of the hauling needs of clients. Foreknowledge of such things as the probability of bumper crops, fuel demands, and the time of the break-up of ice in inland-water shipping regions should permit better allocation of transportation to the benefit of both producer and consumer.

The less developed regions of the world have not been dependent on complex, smooth-running transportation systems. Hence the effects of better weather forecasts on their transportation systems are more difficult to assess. However, these are the areas where major alterations in weather now result in transport blockages. Accurate forecasting could encourage better scheduling procedures, thereby improving over-all transportation.

IMPLICATIONS FOR WATER, FUELS, AND POWER UTILIZATION

It appears that over the next 20 years the main sources of power for most industrial nations will be as they have been: coal, gas, oil, and dammed water. Thus fuel-stockpiling strategies will continue to depend on weather forecasts. With better long-range forecasts, stockpiling could be more efficient than now.

Fresh-water storage could be better planned, a most important implication in view of the growing water shortage. This is especially true of such multipurpose installations as TVA, where some of the reservoir capacity saved to protect against unexpected floods sometimes now means there will be inadequate water stored against unexpected droughts. With a long-range weather forecast it should be possible to allocate water for electrical power and for irrigation more efficiently than is presently possible.

Then too, sudden changes in the weather which produce sudden increases in demand for shipment of bulk fossil fuels can generate lags in the delivery schedule with consequent physical discomfort for the users as well as added expenses if there is a price increase based on the shortage. Coal and oil also could thus be more efficiently stockpiled if there were fore-knowledge of temperature fluctuations. At present, during excessively cold or hot periods there may be sharp increases or decreases in production, which in turn mean either overtime work and pay or reductions in salary. But while smoothing out these perturbations through better predictions may increase profitability for mine owners and oil companies, it may also, if overtime is cut substantially, decrease the take-home pay of the worker and thereby his purchasing power. Thus it is not immediately clear that savings on fuel costs necessarily mean improvements for all concerned. Nor, of course, is it clear that the ultimate consumer would find fuel bills lower. The relationship among the fuel industries and between them and the government are complex, and the economics of the matter are by no means simply related to production and distribution costs.

The problems of planning for fuel demands will probably increase as the demands for power increase with the size of urban populations. Thus the pre-season allocation of fuel and the facilities for transporting it, to the extent that fuel sources and their transportation are international in origin, require international arrangements in the light of forecast weather extremes. The whole philosophy of price, ownership, national policy, and private interests will have to be re-examined in order to insure that fuel shortages or overages do not cause major discomfort or sacrifice for consumer and entrepreneur. The whole complex costing, pricing, and production policies of the American fuel industries and the government will, of course, have to be reconsidered in the light of this international situation.

WEATHER DISASTER MITIGATION

It has been proclaimed that one of the consequences of better forecasts based on satellite data will be weather disaster mitigation. This is true to some extent, but not nearly to the extent claimed. First, let us distinguish between predictions of imminent disasters such as hurricanes, tornadoes, flash floods and typhoons, and predictions of long-range disasters such as droughts and sustained floods. Once this distinction is made, it is clear that a disaster forecast will be useful only if it will allow something to be done in the time left and only if people will do the appropriate things.

Most people are reluctant to take radical action when confronted by threats of disaster. Moreover, because of the infrequency of the threat, the physical situation is often such that not much can be done anyway. Thus, disaster warning information does not *necessarily* produce the desired behavior and the hoped-for saving in life and property unless the predicted disaster is one with which leadership and the people have had previous experiences.

The long-range forecast can eliminate some of the crisis-bred mistakes which occur when there is no foreknowledge of a disaster. It cannot eliminate the need for complex decisions which will offer their own difficulties. Nations might be confronted with grave problems pertaining to population shifts, labor force reallocations, stockpiling of disaster compensating supplies, and the development of control over relocation and recuperation. One can imagine clamorous demands from the population and from the parties out of power for evidences of sufficient advanced preparation. Governments could fall if, after the disaster, there were evidences that more might have been done to get ready for it.

Many countries, of course, do not have the facilities for utilizing such forecasts. Indeed, they would find themselves embarrassed if an impending disaster were made known to their publics. But foreknowledge might encourage other nations to prepare to help mitigate the disaster. These preparations could be altruistically motivated, or they could be part of the efforts to woo the threatened region, to bargain with it for mutual benefit, or to blackmail it into cooperation with the nations having the mitigating resources.

Navigation

The orbiting of the *Transit* satellites has been accompanied by equally high-flying forecasts about the imminence of easily obtainable information, via such satellites, as to precisely where one is under all weather conditions. Satellite proponents point out that conventional methods for determining one's position require relatively frequent rechecking and recalibrating of the equipment and that, in any case, other navigation and position-determining methods do not provide the precise knowledge that navigation satellites could provide.

To assess these claims one needs some understanding about what receiving equipment is required to realize these possibilities. There are two methods for finding position on the basis of navigation satellite information. One method is called sphereographical navigation, which is analogous to taking star sights with traditional transits except that a radio antenna perhaps 10-12 feet in diameter is used rather than a hand-held transit. With such an antenna it is possible to determine position within an error of about a mile when the satellite is about 100 miles away. Smaller antennae would mean larger errors. The other method of using navigation satellite information is called the doppler-shift method, involving the fact that radio frequencies increase as the transmitter moves toward the receiver and decrease as they move away. The difference between frequencies plus some other information, including the precise time, will give the relative position of the observer, which can be translated into latitude and longitude.

However, both systems require revisions in navigation tables as the orbit of the satellite changes—which it inevitably does because of the air drag even at 100 miles and because of other perturbations in its orbit produced by the earth under it. This means, then, that new charts would have to be prepared frequently and provided to all those using navigation satellites for determining their position. Or the satellites would have to be designed to transmit the chart revision data along with the other information they put out.

For the foreseeable future receiving equipment will be bulky or expensive or both. Since commercial ships and planes and ground vehicles can now know where they are with sufficient precision the great proportion of the time, using simpler and cheaper means for navigating, it is hard to conceive of a great peacetime demand for navigation satellites. Ocean liners do not get lost; fishing vessels occasionally do. But it is precisely the ocean liner which could afford the cost of the equipment—and which has the space for a large antenna. (More accurate position-fixing is of course crucially important to undersea or sea-borne missile-firing craft whose weapons depend on inertial guidance.[3]

Space Colonization

When we speak of space colonization, we can distinguish between a colony in orbit such as a small group living in a man-made satellite; a colony in transit, such as a group spending months or years in a space ship to reach other planets; and a colony on the moon or another planet. The distinctions are useful because the technological and social prerequisites for accomplishing one or the other are somewhat different, and thereby also the time to accomplishment. We shall not attempt an exhaustive categorization of the problems for each type of colony. But it is worth noting that the implications of the different types of colonies may be different and so too may be men's attitude toward them.

Probably the most important technological contingency for space colonization beyond the satellite-type of colony is that only when there is nuclear power for space ships can "in-transit" colonies and "on-planet" colonies

---

[3] See Chapter 6.

become significant human activities of more than a few men. Chemically fueled rockets lack the power to transport payloads big enough to make large or long-lived colonies practicable. Even though it will in its more sophisticated versions carry a man around the moon and return him to earth, the giant *Saturn* rocket, in its all-chemical-engine version, cannot land and recover a man from the moon. Some contemplated chemical engines larger than *Saturn* could, in principle, do it, but the efforts usually envisioned when speaking of colonizing a planet appear to require no less than nuclear energy.

If the nuclear power is based on fusion processes—and this would seem to have the most advantages—it would also be usable on earth. And since fusion would mean unlimited cheap power for all peoples, existing political and social organization everywhere would probably be changed, whether or not man lives in significant numbers for significant times on our moon, Mars, or the moons of Jupiter. Such changes would affect the motives and arrangements underlying space colonization. But the spirit and purposes of people having access to unlimited power are difficult to envision, so it is almost impossible to speculate sensibly on the political and social consequences for space colonies built through the thrust of fusion.

However, even if fusion power is not invented in our time, adequate fission-type, upper stage, nuclear rockets should be ready during the next two decades and should provide the basis for some moon landings. (Of course, the technical problems associated with landing on and colonizing another planet are as immense as those surrounding the design of the nuclear engine itself.)

One thing seems quite certain: vehicles to make such trips feasible and the equipment to sustain life on these outposts will not be obtainable by private persons or private groups. Not only would the costs be prohibitive, but the nuclear resources involved presently belong to governments, and there is no reason to assume that this situation will change within our time period. It follows, then, that colonies will exist only if they are in the interest of supporting governments. In our time there will be no interplanetary equivalents to the Mayflower; and colonial revolt on Jupiter's second moon, in a science fiction equivalent of the American Revolution, is far off.

By the end of our time period, whether or not we develop nuclear-powered space craft, we can expect some orbiting satellite colonies containing two to a dozen or more men. But after the glamour wears off and, as with our more routine rocket lauchings, our satellites gradually move to the back pages of the newspapers, these colonizing activities will probably have no special implications. It is more likely that society will treat them as we now treat submariners who stay three months under the sea or the men who winter in the long antarctic night.

However, if men are expected to live in orbit for longer than about six to nine months, they would probably have to be selected in the same way, as discussed below, for on-planet colonies. Thus at some stage the fact of men living in satellites, the by then imminent likelihood if not accomplished fact of two or three men spending many months in space on a round trip (though not a landing) to Mars or Venus, and the likelihood that a few men will have stayed briefly on the moon—all will lead to serious consideration of the social and psychological problems of the constricted life in transit and in on-planet space colonies. The colonists of our time period very probably will be more like the traditional explorers in temperament and behavior. It will be later, if at all, that the social and moral problems will arise. Such problems may never arise for in-transit life if thermonuclear propulsion makes space colonization relatively inexpensive. It would thus be possible to build large and comfortable spacecraft and to take along people necessary not to the tasks of the mission but for the psychological welfare of those who must accomplish those tasks.

In this happy contingency, wives would not need to be scientist-technicians, and crews would not need to be so small that any women among them would generate emotional problems for all concerned under the conditions of isolation and possible danger. If however, space colonization remains expensive, then either mission time will have to be kept short or prevailing standards of interpersonal relationships may have to be modified to allow the colonists adequate emotional fulfillment in their forbidding and demanding environment. It remains to be seen what the effects would be on the rest of society of a different standard of interpersonal relations for space colonists. Perhaps the colonists would be perceived by those "back here" as some sort of elite or at least so different from the rest of us that their social behavior will not be seen as threatening to common moral standards or even as reasonable behavior to emulate.

If space colonization remains tremendously expensive during our time period, effective methods will have to be perfected for selecting crews for these colonies. No nation or group of nations can afford to have a multi-billion dollar space effort ruined by the emotional inadequacy of a crew member. Thus there will have to be tremendous improvement in the predictive quality of psychological tests. Today we cannot even predict good submariners (although we can predict who will do well in submarine school), much less select astronauts with any assurance that our tests are reliable. Although accumulated space crew experience and improvements in written and laboratory selection tests will improve our selection methods, probably these will have to be supplemented by "action screening" for many years. Potential space crews may be screened by having them operate as crews in dangerous situations on this planet.

Simulated danger will not be enough. Exploring the bottom of the ocean in multi-man submarine tractors might be such a training and screening task. Some men will die during these efforts. Furthermore, since the failure or death of a crew member would mean that both his replacement and the so-far-successful crew-members would have to go through the whole experience again to see if the new combination operated effectively as a group, such screening would take long periods of time. In sum, the development of our understanding of human personality will play a large role in determining what can be done in space colonizing.

For these reasons too there will undoubtedly be in the years ahead support for an all-out program to develop paper-and pencil and laboratory screening and testing methods to replace action screening procedures. But better tests will require a deeper understanding of individual and group behavior, and this understanding will doubtless be applied as well to many other facets of organized life "down here." On the optimistic side, one can imagine work groups of all kinds becoming more effective through better meshing of personalities in the group as well as a better tailoring of the group to its mission. On the pessimistic side, one can wonder whether these highly effective selection and screening techniques may not contribute importantly to a state of mind in which it will be believed generally that there is a "right place" for every man.

But man will not be the only factor in the space colonizing system which will have to be pretested in great detail. The variety and complexity of the equipment needed for an adequate biological and psychological life-support system is appallingly great; and all of it must be tested and retested until it is certain that the investment in a colony will not be lost because of a malfunction or operational inadequacy. Moreover, since equipment testing must be done in the environment in which it will be used, and at first at least that environment will be in good part unknown, it cannot be adequately simulated on earth. Certainly equipment which must operate in a weightless condition cannot be tested on earth. Thus many preliminary flights, landings, and extraterrestrial operations will be required to test equipment before men can count on it to support their life and help fulfill their mission.

Among other things, this means that much of the equipment will have to be tested first by remote control, much as we and the Russians have studied the performance of space capsules for men by orbiting them first without men. When this approach is extended to include large satellites, space ship environments for long voyages, and extraterrestrial colonies, we can expect an efflorescence of robot technology on an unprecedented scale.

Two implications of this technology are worth mentioning here. If the robots are very good, it may be easier and cheaper to let them do our space exploration than to do it ourselves. After all, we use machines to extend ourselves into the microcosmic world. Why not use them to extend us into the cosmos off this planet? In the second place, this new technology will arrive just at the time we are to be faced with the full impact of our population explosion. The combination of automation, robotic technology, and too many people would seem more likely to lead to disaster than progress unless we have learned to live meaningfully in a predominantly unemployed society.

There are those who look to space colonies to solve population problems; but, as with so many of our difficulties, this must be solved here rather than in space. For it has been calculated that by the year 2000 if, as is likely, 75 million people will be born each year, it would then take a half million space ships each carrying 150 people leaving one a minute to break even. But there is an ironic twist to this proposal as well. Because of the many physical constraints associated with life in a synthetic environment, those couples who comprise the population of colonies in transit or colonies on other planets will have to be people willing to limit the number of children they have. Thus, to the extent that there is any migration from earth for the foreseeable future, it will be comprised of people who already have the capacity and will to limit their offspring rather than of those who are already overpopulating the planet!

Since there has been some speculation about exploiting the resources of other planets for the benefit of this one, it is worthwhile to examine this fantasy. Certainly within our time period it will not be practical, for example, to "mine" the moon. All planets have the same elemental composition, and any resources that can be found there can also be found here. The only differences which could be potentially significant are the amounts of the raw product and its purity in its extraterrestrial form. So far, however, no one has imagined any naturally occurring product so valuable that it would be worth transhipping to this planet. This is especially so given our growing capacity to invent synthetic products tailored to the particular needs of society. Moreover, the ocean contains practically unlimited supplies of all the elements. Once the day of fusion power is here, it would be much easier to extract minerals from the ocean than it would be to mine, purify, and tranship them from anywhere off this planet. Although someday the raw products of another planet might well be mined for the use of colonies on that planet, that day is not coming soon.

Rocket Transport

Will the day come when rocket transport will be a common means for moving men and material on peacetime tasks? To judge by frequent forecasts, the day, while not around the next corner, is claimed to be certainly not far down that block. Then, it is confidently predicted, rocket ships will carry men and material from space ports modeled more or less after today's airports to all points of the globe within a matter of a few hours at most. Two types of rocket transport are envisioned. On the one hand there would be more or less conventional rockets, huge enough to carry economically profitable payloads, blasting off as at Cape Canaveral today and either gliding to a landing or being parachuted to earth or letting down on retrorockets, in a maneuver which is the reverse of launching. On the other hand, there is envisioned a piggyback arrangement whereby the rocket and its payload are lifted off the ground first on the back of a jet plane and then released at an appropriate altitude to complete the flight as a real rocket which must also reenter the atmosphere and land under its own control. Either way, high speed monorails or some such system is foreseen speeding traveler or freight from distant rocket fields to their ultimate destination.

Is this fantasy one likely to become reality in the next two decades? Consider first the requirements for moving freight by rocket. The products to be moved would have to be either designed or packaged to withstand rocket accelerations and landing decelerations. They would have to fit the payload compartment of the rocket or be disassembled in pieces which would fit. Because of the noise and danger, rockets are not going to be launched as close to town as planes now fly from airports. Nor will they land close to town. Since special high speed transportation to major airports has seldom been economically feasible, it is doubtful that it would be profitable for space ports either, at least as far ahead as we can see. Thus the item to be lofted into the upper reaches of the atmosphere would first be carted mundanely to the outer reaches of local civilization.

We can ask, then, what products in the day of the jet freighter—which certainly must come at least as soon as the freight rocket—are worth specially designing, specially packaging, and spending all the time to tranship to and from rocket fields just to cover the intervening distance at a quarter or a tenth of the speed of jets. We are hard pressed to think of any. Perhaps rockets could be used in emergencies, but emergencies are hardly a basis for operating a commercially practicable transportation system. And it would be a fortunate coincidence indeed if the objects in need of rapid transportation were also ones which could be safely transported by rocket, and that launching and recovery areas would be where needed when needed for emergency.

Other items in the day-to-day intercourse of men which need to be transported swiftly can be planned for ahead of the time they are needed —especially as automated inventories and planning schedules become the rule. But to the extent that one can plan ahead for these items they can be moved by simpler, safer, and probably cheaper transportation forms.

There have been a few demonstrations of the feasibility of sending mail by rocket. It can be done, to be sure, but there is a real question of its practicality. Mail rockets can no more land in or near cities than can other rockets. The microminiaturized letters which would travel this way would take equally little room on a jet or supersonic jet transport. It is hard to imagine a regular demand for a service whose only special asset would be saving a few hours in delivery time, especially if inexpensive international telephony relayed by satellite were also ready when minutes count. Wherever a printed document would be needed fast, facsimile reproduction via satellite or ground cable might be more economical.

Now let us consider the movement of men by rockets. Here, presumably, the incentive which might make this effort commercially feasible would be that of getting somewhere far away faster than by jet or, within the next two decades, by supersonic jet. Here, of course, packaging problems will arise which will be more complex than for inanimate objects. Also, safety standards must be much higher when people are the payload. Given the amount of additional equipment which would be required to package people safely, the profit margin for such an enterprise might be slim indeed—and certainly, as far as we can look ahead, rockets will not be as comfortable as jet.

To be sure, there will undoubtedly be situations where men would still want to travel faster than jets permit, but then we wonder whether the enlarged television and telephone capacity at this future date made possible by communication satellites will not in crucial circumstances take the place of actual physical transportation. Certainly, it would be cheaper and safer. Finally, such high-speed transportation would present the commuter with a time problem similar to that which will confront live, transcontinental, and transoceanic TV telecasts. The three-hour difference between New York and California becomes five hours difference between New York and London: lifting off at a reasonable hour in one place does not necessarily land the commuter at a reasonable hour in another.

Thus, for the foreseeable future, freight transportation—man or machine—does not appear to be a practical commercial peacetime use for space activities.

Space as a Research Tool

Space will provide an environment for research as well. Fundamental discoveries will be made in biology, astronomy, and geophysics. In fact, there have already been major discoveries in geophysics and astrophysics, such as the Van Allen radiation belts, and some evidences of the persistent radiation pressure millions of miles from the sun called the "solar wind." And important discoveries in engineering science will certainly be made in space.

One should not assume that space research is all that is needed to advance science. On the contrary, as with meteorological research, other types of space research are necessary but not sufficient conditions for optimally advancing any area of science. Many space enthusiasts have forgotten or ignore this fact in their single-minded devotion to the rocket payload. And many scientists, even including astronomers—who, it is popularly believed, will find the greatest research bonus in space probes— are most unhappy about the evidences of unbalanced research programs. Great efforts and funds are poured into anything which leaves the atmosphere, while the equally important projects down here have trouble finding adequate support and are sometimes left undone. The problem of allocating a balanced effort among the sciences and within a given science is indeed vexing and subtle. But the reader interested in the research potentialities of space cannot afford to ignore it. For within this problem lie many of the pressures and much of the politics which will determine which areas of knowledge will benefit the most and which the least from space research.

What follows will be directed mainly to the basic research possibilities of space; the applied research being done and to be done is endless. Since no one can predict the discoveries which will flow from basic studies or forecast specific uses for these discoveries, we shall not attempt footless fantasies of this sort even though they are regularly indulged in by the public relations camp followers who besiege every space effort. It will be evident that some research programs, especially in biology and psychology, will have general implications for life here on earth, and these will be pointed out in passing.

BIOLOGICAL RESEARCH

While many aspects of the space environment can be more or less simulated on earth—such factors as radiation, vacuum, temperature—the effects on biological organisms of a weightless environment can be explored and understood only in space. Many of the gross physiological processes are believed to be able to operate effectively in a weightfree environment; although no formal data have yet been released, the Russians claim that at least over periods of a day or more organisms can carry on the physiological processes needed for survival. The effects of longer exposure are not yet known.

On the biological level of research, as contrasted to the physiological, there are many basic studies to be done on the effects of a weight-free environment on shape and structure in growing biological organisms. We know that increasing the effects of gravity, as can be done by raising organisms in a centrifuge, has some effects on shape and growth; and Russian experiments indicate that cell division is faster in space. Scientists strongly suspect that a lack of gravity will have many other fundamental effects on basic biological processes. Understanding these effects in the light of our knowledge about growth and structure on earth will add vastly to our understanding of the nature of life.

But there is another environment off this planet even more exciting for biological research than the weightless one of space itself. If life in any form should be found on other planets—and there is reason to believe that primitive lichen-like forms exist on Mars—an understanding of its forms and processes in comparison with ours should provide biology with the kind of major insights into life processes which can be obtained only when old perspectives are exposed to a new environment. The advances in biological theory so stimulated should be profound.

Research on man in the space environment should contribute special knowledge about ourselves which, for ethical and practical reasons, we have not obtained on earth—namely, an understanding of the processes involved in coping simultaneously with extreme psychological and physiological stress. Although we know that man can sometimes show extraordinary capacities under extreme stress, we know very little about what, in fact, is involved in this. For as far ahead as we can see, putting man in space will mean great risk. One of the requirements for participating will be willingness of the astronauts to serve as guinea pigs by permitting as much as is technically possible of their physiological and psychological states to be recorded and transmitted to earth for analysis. Thus, there will be records to help us understand better what has happened to the man and what he does—or fails to do—in the event of extreme stress.

Then too, in creating means for man to cope with boredom, space-radiation, and other experiences associated with long space flights, new drugs are likely to be found. These will give further understanding of man under medication in special environments—knowledge useful on earth as well as in space.

Space has stimulated the development of subsidiary research techniques, one of the most important of which, now refined beyond its earlier forms, is biomedical telemetry. This is a method by which it is possible to sense via miniaturized electromechanical devices a multitude of physiological signals and transmit them to a remote receiver where they can be recorded and analyzed. Such telemetry techniques no doubt will be further refined, and they will lead to a better understanding of other biological, physiological, and psychological processes on earth and in space.

PSYCHOLOGICAL RESEARCH

It will be worthwhile sending man into space on tremendously expensive missions only if he is uniquely capable of performing tasks which machines cannot perform adequately. Such tasks will require judgment, since any motor or sensory skill can be reproduced or improved upon in a machine. But tasks of judgment are intimately bound up with the personality of the judger and with the social situation in which he finds himself. Thus, if humans are to have the required capacity for judgment, more knowledge will be necessary about the effects on judgments of personality and the social environment.

Research on the effects of boredom and of continuous association with the same small group of people over long periods of time will be central. Whalers, submariners, and polar explorers have spent much time together, but the circumstances are not precisely analogous in that the working environment, the size of the group, the tasks to be performed, and the backgrounds of the men appear, in one way or another, to be different from what might be important in space crew operation. Thus, the research based on the requirements for highly competent manpower, as well as the experiences these people have in space, should give us new knowledge about interpersonal relationships and the relationship between the emotional characteristics of man and his competences as a skilled rational creature operating in a group.

GEOPHYSICAL RESEARCH

Satellites and space probes offer, and indeed have already demonstrated, special possibilities for helping us to understand better our own planet. We have discussed the opportunities for fundamental research in meteorology. Similar opportunities exist for studying gravity, the earth's electromagnetic fields, ice drift patterns, heat radiation and absorption, and a variety of other phenomena having to do with our planet. Photographic surveys covering tremendous surfaces of the earth should help us to refine our knowledge of the earth's geography as well as the relationship between its surface geology and various manifestations of gravity and electro-magnetic fields.

ENGINEERING RESEARCH

The space environment should offer unique opportunities for research in areas new to engineering science. The accessibility in unlimited amounts —if one can speak of vacuums this way—of better vacuums than can be produced in the laboratory will foster the study of new fabricating techniques and manufacturing processes which depend on uncontaminated and/or gas-free environments. Whole new large-scale technologies may be developed, for example, for vacuum distillation in orbiting "factories." This vacuum, in conjunction with very high temperatures, produced by focusing the sun's rays, and in conjuction with the very low temperatures of space in the shadow of a satellite, may stimulate methods which would make it worthwhile transhipping certain products from the ground to orbit and back again.

Whether these possibilities will be realized depends on what research discovers about the conduct of engineering in space—in particular about the design of engineering systems which will operate in a weightless environment. All systems which now operate by subtle uses of gravity will have to be reanalyzed and redesigned so that they operate in a weight-free environment or even independently of whatever direction gravity might impinge upon them.

It hardly needs adding that the extremes of the space environment will, as they already have, stimulate basic research in all those engineering areas which space craft depend upon for their existence. Metals, lubricants, structures, plastics, instrument design, to name only a few, all will require extensive research programs if the technological wherewithal to meet our aspirations is to be realized.

ASTRONOMICAL AND ASTROPHYSICAL RESEARCH

Astronomers have described their earth-bound "seeing" difficulties as being worse than those confronting a man under several feet of water who is trying to discern the nature of an environment above the waves which he has never seen in any other way. The mantle of air which makes life possible also makes it very frustrating for astronomers. For that air absorbs, distorts, refracts, and reflects most of the frequencies of electro-magnetic radiation which are crucial for extending our understanding of the basic processes of the universe—indeed, of our very sun. Our seeing ability even within the range of wave lengths to which our eyes are sensitive is poor. Our ability to "see" in the very long and the very short wave lengths in space is not only poor; over large parts of the electro-magnetic spectrum it is nonexistent: the difficulties produced by the atmosphere are compounded by the effects of the ionosphere and our earth's magnetic field.

Thus much of the fundamental knowledge that astronomy needs for understanding the great cosmic processes cannot be gained as long as we are kept within our atmospheric mantle and, for some research purposes, as long as we are within the general environment of the planet which extends far beyond the atmosphere. Such knowledge will require instrumented probes high above the atmosphere. To be sure, such research programs are already underway. These will be enlarged and evermore rewarding as we are able to launch heavier payloads and more sophisticated equipment to operate in deep space or from the surface of airless planets.

Attitudes and Values

Any essay speculating on peaceful space activities would be incomplete if it did not examine the effect of space on the values and attitudes which undergird and justify the things men find important, the goals they aspire to, and the social forms by which they fulfill or frustrate themselves. Since the commitments men will make to space activities are themselves expressions of more fundamental values and attitudes, an examination of the general relationships between values, attitudes, and space activities becomes doubly pertinent.

So far in this chapter we have drawn attention to likely shifts in values and attitudes as a result of the application of particular space activities, for example, the probable effects of satellite-based weather forecasts on the attitudes of farmers and their governments in underdeveloped nations. In this section we shall speculate on the possible effects on peoples' attitudes and values of space activities as a more generalized endeavor.

Each space development or potentiality will inspire people or leave them indifferent, and will be understood or misunderstood, to the extent that it meshes with the prevailing world view of the spectator or participant. There is ample evidence that man is conservative when it comes to changing his personal world view, our frequent ritual mouthings about wanting progress notwithstanding (as any examination of what we do and do not do with most of the potentialities of our technology will verify). Most people attend carefully only to experiences which are immediately significant in their own everyday life. They react to new experiences in terms of their learned and tested mode of responding to the world. They perceive in terms of their preexisting values and attitudes; they try to mold new experiences into old contexts; and thus the new experiences frequently lose their unique implications and power. If the experiences do not fit this standard context they are likely to be ignored altogether. The experience of changing attitudes and values is an upsetting one that most people try hard to avoid.

Nevertheless, it usually is asserted that space exploration will broaden man's perspectives and his view of himself and thereby make a finer creature of him and a better society for his fellows. If space activities do come to directly touch a good part of society, they doubtless will affect values and attitudes. Until that day it is likely that for most people what happens in connection with space will reinforce old values and attitudes rather than stimulate new ones. For the most part, men will continue to look at the new horizon of space through old eyes—when they are not staring at some different horizon altogether.

Indeed, one need only look at the things presently emphasized and distorted regarding space activities to recognize that so far space has been used chiefly to grind old axes finer. Consider, as examples, our attitudes about space competition, prestige, and booster size; the intense battles over the merits of one governmental, organizational, or corporation approach to the future of space compared to another; the confusion of science with engineering and scientists with science administrators; and, above all, assertions about the pros and cons of "national interest." All of these are almost invariably expressions of deeper values and attitudes about how space should be used to advance the special interests, commitments, and aspirations of the particular proponents of one set of criteria for the use of human resources rather than another. And these interests, commitments, and aspirations are, of course, established ones. As such, for a long time they will be not so much changed by space as space will be molded by them.

How then are space activities supposedly to affect man's values and attitudes? Five arguments are common: Space activities will broaden man's horizons, be a major if not the major piece in the prestige game, enhance the role of learning and science, replace war, and improve international relations.

On the first of these, we have already said enough except to point out that we do not know what the effects of space activities will be on the attitudes and values of today's children who will be tomorrow's leaders and taxpayers. For some of them space undoubtedly has a reality it cannot have for most adults, with their lifetime of pre-space perceptions. But it is not at all clear that playing space-man today means more interest in and support of space activities twenty years from now. It might; but we know too little about how values change from generation to generation in a rapidly changing society to predict strongly one way or the other. Every boy who tinkered with a Model A did not grow up interested in engineering, or to be an engineer, or, if he did, to share similar values with other engineers.

Judging from the invariable equation of space with prestige, one might expect that fundamental shifts in future attitudes toward the major space contenders might be produced by what they do with their rockets. Here is a classic example of space reinforcing old preconceptions about the nature and effects of public opinion. But we do not know enough about "prestige," nor have we sufficiently defined the values and attitudes presumably associated with it, to know whether or not it *is* affected by specific space activities. Do we mean by prestige that, if we have it, people will like us, or emulate us, or send their students to our schools, or buy their heavy equipment and instruments from us, or feel that we would win a war if a showdown came, or turn their back on offers and blandishments from the other side—or all of these?

But more than this, data from here and abroad do not show a clear relationship between the attitudes held toward a nation and its space accomplishments—once one goes beyond the irretrievable transformation wrought by *Sputnik* in the image of Russia into that of a first class technological-military power. Beyond this shift, it has not been shown that major changes in views of the United States and Russia have been due to subsequent space accomplishments as such. In fact, it is possible to show that a good part of the world's populations, including Americans, knows and cares little about space activities.

There is no good reason to believe that as long as space is a competitive area rather than a cooperative one there will be anything either contender can do technologically which the other could not do eventually or sooner if he wished to make the effort. And there is no reason to suppose that this fact is lost to decision makers and leaders everywhere. Thus in the peaceful areas of space it is unlikely that accomplishment *per se* at a given time will produce major and permanent shifts in attitudes toward either of the contenders.

Most countries of the world do not have to be reminded how central to their growth are education in general and science in particular. For many people in the United States this dominant fact was not evident until *Sputnik*. It still evades many of them or is perceived only in the context of East-West relationships. Many of those who do take space and education seriously, have a tendency to give space the credit for our new, even though limited, awareness of the priority of good education. One suspects, however, that if the Soviet Union had come up with a form of wheat that grows to maturity in two weeks, the effects on education would have been the same in this country. Space instead was the accidental key to the back door of the schoolhouse, but, once in, space also makes a convenient wedge for keeping the door open. In this sense, it may be helping to change attitudes and values about the *utility* of education—whether for beating the Russians or for getting a high-paying and glamorous engineering job. It is not at all clear, however, that space is generating by itself an attitude toward education as a good in itself, with an appreciation of the value of non-utilitarian learning.

Certainly *Sputnik* also opened the door for the scientist to high places in government; it gave a permanent role to scientists as valued and powerful advisors, and, through their advice, inadvertently or deliberately the role of policy makers. Some scientists and non-scientist policy makers are prepared to acknowledge that this new relationship poses problems for both groups as to their legitimate professional roles. Others in both groups sanguinely assert that the prevailing values and attitudes are adequate and undisturbed by the new arrangement of power. But in the nature of power and in the devices used for its application, both groups will inevitably find their values and attitudes changed—whether for the good or bad of science and government is yet to be determined. Space activities will provide some of the pressures which will mold these changes. However, here particularly, it is unlikely in our time period that space will play nearly as crucial and determining a role as it did by first bringing science and government so closely together in the days of *Sputnik*.

A frequent, almost wistful, speculation on the true value of an "all-out" space exploration effort is that it could take the place of war because it would siphon off the aggressions of man by transforming them into the commitment and massive effort needed to conquer space. What little we know about the psychology and sociology of man which leads him to make war on his fellow man indicates that this substitution theory is so naive as to be only a charming but footless exercise in wishful thinking.

This is not to say, however, that space activities may not importantly contribute to a change in values which may help reduce the tensions and distrust which contribute to an unstable world situation. Weather forecasting, world-wide communications, and really big man-in-space colonization efforts will require the cooperation of many nations. Not only are costs and resource demands likely to be too great to be borne by single nations, but some space activities simply will be useless without other nations. The reconciliation of conflicting goals and means for reaching them, and the opportunities for scientists, engineers, and professionals of many kinds together to work out or live with these differences, may make it easier for decision makers, opinion leaders, and even the ubiquitous man in the street to accept the idea of one world.

Since the resolution of problems related to space activities will require intensive examination of many of our values and attitudes about problems so far unrelated to space, it is not impossible that space will provide a slow and quiet leverage for prying us loose from ideas more or less successful in the simpler world of yesterday but incompatible with the intimately intermeshed world of tomorrow's technology.

On the other hand, it is not certain that space activities will point the way to greater sanity: the tail may not wag the dog. It is all too easy—especially when seduced by glamorous predictions—to equate escape from Earth by rocket with escape from Earth's present and future problems. In most of us is the fantastic wish for a good fairy who will make everything all right, and now the good fairy wears a space suit. But the military uses of weather information and communications, and the frantic arguments of some that we must control the "high ground" and gain "control of space" in order to control earth, may mean that our present militaristic, nationalistic values and attitudes will be reinforced, and that space rather than leading us to far horizons will simply bring the clouds of fallout closer.

This chapter should have demonstrated that there are wonderful uses for space if we choose and use them sensibly. We can hope that outer space will help us to develop the appropriate attitudes and values. But, finally, it is the inner space of man where the decisions are made and the actions initiated which will determine the uses to which space is put. Looking up is inspiring and can help us to look *in,* which in these days is terrifying. But looking in must come first if we truly hope to go on looking up.

3.

The impact on the
American economy

♦ LEONARD S. SILK

Three great events stand at the threshold of the modern age and determine its character: the discovery of America and the ensuing exploration of the whole earth; the Reformation, which by expropriating ecclesiastical and monastic possessions started the twofold process of individual expropriation and the accumulation of social wealth; the invention of the telescope and the development of a new science that considers the nature of the earth from the viewpoint of the universe. . . . In the eyes of their contemporaries, the most spectacular of these events must have been the discoveries of unheard-of continents and undreamed-of oceans; the most disturbing might have been the Reformation's irremediable split of Western Christianity. . . . However, if we could measure the momentum of history as we measure natural processes, we might find that what originally had the least noticeable impact, man's first tentative steps toward the discovery of the universe, has constantly increased in momentousness as well as speed until it has eclipsed not only the enlargement of the earth's surface, which found its final limitation only in the limitations of the globe itself, but also the still apparently limitless economic accumulation process.

Hannah Arendt
The Human Condition

◆ LEONARD S. SILK is Senior and Economics Editor of *Business Week* magazine. A onetime professor of economics, he served the U.S. government both in Washington and in Paris during the early 1950's. He received the Loeb Award for Distinguished Business and Financial Journalism in 1961, in addition to earlier fellowships. His books include *The Research Revolution* and *Forecasting Business Trends*.

The Cost Factor

Even if space programs were nothing more than an industrial civilization's equivalent of pyramid-building, the magnitude of the present effort would be impressive. And we are just at the beginning of really massive financial outlays for outer space.

From fiscal year 1955 through fiscal 1961 the United States government appropriated some $4½ billion for various space programs of the National Aeronautics and Space Administration (NASA), the Department of Defense, the Atomic Energy Commission, the National Science Foundation, and the Weather Bureau. In fiscal 1962 those appropriations increased by an additional $3 billion, to bring the cumulative total to about $7½ billion through July of 1962.

These figures are based on the minimal definition used by the Department of Defense to measure its own space activities—a definition which includes only those amounts spent by the military services for certain designated space and space-related projects (such as *Vanguard, Explorer, Discoverer, Midas, Samos, Transit, Advent, Courier, Space Track, Spasur, Blue Scout,* and the *X-15*). They do not include substantial amounts for construction and operation of the national missile ranges for use in the space programs; the cost of developing missiles, such as *Thor* and *Atlas,* which are also used in space programs; or supporting research and development (such as bio-medical research) which is more or less mutually applicable to other programs also.

If adequate account is taken of those excluded—but essential—Defense expenditures, the total outlay on space by the Defense Department would be increased from the fiscal year 1962 level of $1 billion to a figure double, or possibly treble, that amount. Some analysts place the true Defense Department space outlay at approximately $2.3 billion for fiscal 1962. This more realistic figure, when added to NASA's fiscal $1.7 billion and other agencies' smaller expenditures, implies that in fiscal 1962 we shall spend about $4.2 billion on all military and civilian space programs. (See Appendix A.)

It is extremely difficult to project how high space expenditures will go from here, given all the political, military, and scientific complexities. Even before the Soviets' successful earth orbit by Majors Gagarin and Titov and America's successful suborbital flights by Commander Shepard and Captain Grissom—and President Kennedy's decision to step up the United States space effort—it appeared to many observers that substantial increases in space spending would occur throughout the 1960's. On October 7, 1960 one experienced industrial economist, Murray L. Weidenbaum, estimated that total NASA expenditures for the 10-year period through 1970 would amount to $17½ billion, mainly for big booster programs such as Saturn and Nova and vehicle and payload development.[1] Weidenbaum estimated that NASA expenditures would rise from about $500 million in 1960 to $2.2 billion by 1967 and thereafter level off. His NASA estimates now look too conservative. Although he did not break out projections of military expenditures on space and space-related programs, Weidenbaum was clearly expecting a continuing rise in that area of defense expenditures.

In January of 1961 another industrial analyst, D. C. Eaton, estimated that major military procurement for space would rise from an annual rate of about $900 million in 1960 to $4 billion in 1965 and to $6.5 billion in 1970. Eaton's estimates imply that major procurement for space programs in the decade will amount to $40 or $50 billion. From the military standpoint, Eaton argued, "our whole defense will rest upon the improved ability of our space systems. . . ." Over and above the strict military requirements for space systems, he assumed growing outlays on space exploration and the development of navigation, communication, and

---

[1] M. L. Weidenbaum, "The Changing Structure of the Military Market," Conference on Space Age Technology, American Management Association, New York, October 7, 1960.

weather systems. "Should there be favorable changes in the international situation," Eaton predicted, "the proportion of monies devoted to the exploration of space will be even greater by 1970."

NASA has no official estimate of the cost of space programs for the next ten years; the rush of events has made earlier guesses obsolete. Before the Kennedy decision on the moon mission some NASA officials were talking in terms of a 10-year program costing $25 to $30 billion—"although $40 billion could be right." These estimates were admittedly very rough—and stated in terms of what 1961 dollars could buy in this area. But NASA officials have themselves estimated the cost of a three-man mission to the moon alone at $20 to $40 billion. An acceleration of that program would obviously swell the bill for the 1960's.

Some analysts believe actual space outlays are likely greatly to outrun any predictions government officials have thus far been willing to make. James Williams, of United Research, Inc., notes that major research and development efforts conducted by the Defense Department and NASA and other agencies have typically overrun original estimates by a factor of 2-3 times. Williams therefore takes such projections as the recently-anticipated $20 billion outlay by NASA in the 1960's plus the $30 billion which some consider reasonable projections of Defense Department outlays on space in the next ten years and multiplies them by 2 or 3 to get a total of $100 billion to $150 billion for the outlay on space programs over the next decade. This sum may sound unbelievably high. Williams notes, however, on the assumption that total defense outlays over that period amounted to $400 to $500 billion (assuming that the cold war continued in something resembling its recent state, and that defense spending continued at its present level, or increased only moderately), that outlays of this magnitude would constitute only about a fifth to a third of total defense expenditures over the next ten years. A greater emphasis on conventional or paramilitary defense, which the Kennedy administration has advocated and which some observers anticipate, would push the total defense bill up markedly higher. So, obviously, would price inflation.

Admittedly, there is some possibility that such projections for NASA and Defense Department space programs over the next ten years may be too high. There may be a heating up of the cold war and a diversion of funds (and of human and physical resources) away from space programs to more immediate military uses; or a shift in Soviet "competitive coexistence" strategy away from the space area to some other area which we might be moved to follow. We may have a national election in which a political party with a highly conservative budget and fiscal policy receives a national mandate to cut back federal outlays. Such developments could cause the actual expenditure on space during the next decade to drop well below the totals presently foreseen by industrial or government economists in the space field, whose vision may conceivably be distorted by their organizational interest in seeing space programs expand.

Moreover, although there is certainly a considerable degree of public enthusiasm for a space effort of some size, and a desire "to beat the Russians," it is far from clear how large an outlay for space the taxpayers would be willing to support. The first Gallup poll on this subject, published May 31, 1961, found that only 33 per cent of those polled said *Yes* to spending $40 billion to get a man on the moon, 58 per cent said *No,* and 9 per cent had *no opinion.* There are also signs that some scientists and other informed observers consider that expenditures of the magnitudes presently contemplated would be an unwise use of resources compared to other scientific, welfare, military, or foreign aid programs that could be undertaken. For example:

> Rather let us make a balanced scientific effort, including space research, but informing the world that we are more interested in other things than racing to the moon. Let us improve our medical research and make it available to the world. Let us put more effort into social, economic and political research, to be used to ameliorate and improve the human condition at home and abroad.
>
> Let us encourage cultural development—literature, the arts, music—at home and abroad, as evidence that we do not subscribe to the Communist idea that man is essentially an economic automaton. Let us improve our foreign aid. Let us keep up our defenses. Let us improve our educational system for all our citizens.
>
> Let us strive harder to realize our ideals of human rights and equality under democracy.
>
> These are the things on which we could well spend $20 to $40 billion instead of shooting it to the moon.[2]

But even the skeptics rarely go so far as to suggest junking space research or neglecting programs for the military utilization of the space environment for defensive or arms control purposes. And there is such evidence in favor of increased expenditures as that furnished by the *Harvard Business Review* poll [3] of business executives, who declared that they preferred expanded space programs even to tax reduction, and by Congressional and press reaction both to the successful suborbital *Mercury* shots and to Mr. Kennedy's space message.

Thus, although the crystal ball is very clouded as one peers into space, it appears probable that the American people are likely to favor and to back with their tax dollars a greatly increased space effort in the decade ahead. It seems safe to predict that space programs running into the tens

---

[2] Letter to *The New York Times,* May 28, 1961, by Herbert S. Bailey, Jr.
[3] Issue of September-October, 1960.

of billions of dollars over the next decade will reach into nearly every channel of American business, and that space will constitute a market which no businessman or investor will want to ignore.

The Complexity of Industry's Role

To get some notion of the widely-ramifying effects of the space effort on American business, consider the contract structure of a single past venture, the U.S. Navy-directed *Vanguard* project for orbiting a four-pound six-inch diameter, "sophisticated" (a lovely space-age term of technical approbation) satellite. *Vanguard* had a miserably black day on December 8, 1957, when it was supposed to provide America's first answer to the big Soviet *Sputniks* I and II, both already in orbit. That day—at 11:44.559 in the morning—the *Vanguard* space vehicle blew up on its pad, and the sophisticated grapefruit, still beeping away, fell into the flames. But *Vanguard* went on to achieve some brilliant successes which contributed much to the discovery of the Van Allen radiation belts and gave the United States its first tracking system.

Who had a hand in *Vanguard?* The Martin Co. was the prime contractor and was itself responsible for the background *Viking* project, the airframe of the first stage, the nose cone, the third stage support, spin and bearing, and certain ground support equipment. The General Electric Co. provided the first stage propulsion; Aerojet General Corp. handled the second stage propulsion; and Grand Central Rocket Co. and Hercules Powder Co. the third stage propulsion. Atlantic Research Corp. built the spin and retro rockets and the igniter. Air Associates, Inc. made the accelerometers, the Empire Devices Corp. the noise intensity meter, and Hoover Electronics Co. the frequency converters. Henry O. Berman, Inc. made and installed the control equipment. Designers for Industry, Inc. did the programmers. And Minneapolis-Honeywell Co. built the three-axis gyro reference system.

The Behnson-Lehner Co. supplied the data reduction section equipment for the Martin Co.; and Reeves Instrument provided the computers for the *Vanguard* section. The Ludwig Henald Mfg. Co. built the instrumentation and transportation trucks for the ground handling of data; and Polarad Electronics supplied the ground handling test equipment. Loewy Hydro-press built the static firing and launching structure, Derby Steel Co. the test stands and related tools, and William T. Lyons Co., Inc. the test stand for vertical assembly at the Martin Co.

Dian Laboratory, Inc. computed the longitudinal vibration of the vehicle; and the University of Maryland handled the wind tunnel tests (to determine from small models what air drag and air pressure and heat the vehicle would feel along its skin as it moved through the atmosphere). Bendix Radio was responsible for the Minitrack system for spotting the satellite's orbit. Radiation, Inc. supplied the data recording and reduction equipment. Elsin Electronics constructed the telemetering ground stations. International Business Machines handled the orbit computations. Brooks & Perkins Co. built the shell and braces of the satellite's structure. The Raymond Engineering Co. made the satellite's separation mechanism. The Connecticut Telephone & Electric Co. made the transistorized decoder. And dozens of other companies and nonprofit institutions—such as General Radio, Fruehauf, Hewlett-Packard, Tektronix, Melpar, Hazeltine Electronics, New Mexico College—supplied beacons, antennas, and miscellaneous units and equipment.

Moreover, a full accounting of U.S. industry's contribution to *Vanguard* would also have to include many other companies that supplied materials or equipment or transportation or other services to the project's prime contractor and subcontractors and also to the Navy (especially through the Naval Research Laboratory, which managed the entire venture) and to the Army and Air Force, both of which were ultimately involved— the Army by building and operating the South American Minitrack stations and the Air Force by contributing the missile firing range and by plotting and relaying some of the *Vanguard* satellite's data.

The national effort behind that four-pound, six-inch, long-lived grapefruit-in-space (the *Vanguard I,* with its high orbit, is expected to continue swinging around the earth for a thousand years) is fantastic to contemplate.

Yet the *Vanguard* project was a small venture compared with the total space program today, and even smaller compared with the billion-dollar programs that lie ahead. When *Vanguard* was first announced back in 1955, it was estimated to cost $20-$30 million. The project wound up costing (as of January 1959) $111,085,000, or roughly four times the original estimates. The *Mercury* project, originally estimated to cost $200 million, will probably wind up costing about $500 million by the end of 1961.

SOME SOCIAL PERSPECTIVES

It is difficult enough to maintain perspective on outlays amounting to $100 million, let alone on outlays in the many-billions class. *Vanguard* alone involved expenditures greater than many large industries will spend on their research and development programs in 1961—stone, clay, and glass ($111 million), rubber products ($109 million), paper and allied products ($93 million), textiles and apparel ($47 million). One hundred million dollars would pay for half a million air-conditioning units. It would cover the annual college tuition fees of a hundred thousand students. *Vanguard* cost more than 25 times as much as the estimated costs of such other highly regarded research projects as Table Mountain in Colorado for studying the earth's atmosphere. If the marginal value for the nation of space expenditures in the $100 million class requires strong justification when measured against the types of alternate uses of funds suggested above, consider the difficulty of providing the justification for space outlays in the $50 billion class.

Let us not ignore the possibility that in a somewhat sated, high-consumption society such huge outlays might be justified simply for their employment or income effects—a form of sophisticated pyramid-building. Consciously or unconsciously, they might be favored as a means of providing a cure for economic stagnation or as an alternative use of resources should defense programs achieve levels of "overkill" capacity of "stable mutual deterrence" where military procurement could be markedly reduced.

The latter possibility—the reduction of military procurement, to the detriment of the industries involved—raises an issue worth noting here.

President Eisenhower in his Farewell Address suggested that some defense contractors were engaged in efforts to get the government to maintain military expenditures at extraordinarily and unnecessarily high levels for reasons of vested interest; he was apparently making the type of charge leveled by Lord Byron in the *Age of Bronze* against the landed English country gentlemen after the Napoleonic wars:

> The last to bid the cry of warfare cease,
> The first to make a malady of peace.

Can such charges be made against those who are advocating enormous expenditures on outer space projects? Certainly the charge that space is essentially a massive boondoggle would be bitterly resented by the companies involved. It would be countered by the contention that United States private industry, the most resourceful and inventive in the world, through its participation in both space and defense programs, is contributing enormously to the strength, security, prestige, welfare, and growth of the nation. Many companies heavily involved in defense programs refuse to grant the contention that their survival or growth depends upon a continuing supply of federal contracts, or to accept the implied charge that they are "merchants of death." One major diversified defense supplier states his company's philosophy this way:

> Reviewing the history of two almost disastrous wars and appraising the technological and ideological realities of this century, it is my conviction that defense must be a permanent business, a permanent way of life. And that in order to keep in being those aggregates of vital skills that underlie our capability not only to be strong but always to appear to be strong, the modern industrial corporation must, I believe, have a broader base. It must be strong in defense production, strong in civilian production . . .

Not all companies with a stake in the space or defense business, however, would be ready—or able—to survive a drastic reduction in defense or defense-type expenditures. Since very heavy space expenditures are justified by their proponents as essential to national prestige in the cold war, or, more directly, for their relation to defense or arms control, they may be regarded in large measure as defense-type expenditures.

The Prospects for Industry

However one evaluates the admixture of national and self-interest on the part of companies already involved or wanting to become involved in the space effort, the objective fact is that the venture into space looks extremely attractive to many areas of American industry. It will have significant effects upon profits and growth of the economy, upon the structure and skills of the labor force, upon the composition of national resource and capital equipment requirements, upon the level of research and development expenditures, and upon the geographical location of industry.

In our sketch of the contract structure of *Vanguard* we noted how that project cut across many industry lines, involving dozens of widely scattered companies. A comprehensive view of the entire NASA program would of course fan the distribution of contracts out still more widely.[4] It may be noted that all of the companies with substantial NASA contracts also stand high on the list of Defense Department prime contractors for experimental, development, test, and research work. As space programs expand, it is safe to say that many other companies and nonprofit institutions heavily involved in Defense R & D will turn up on NASA's list of substantial contract holders, probably augmented by some companies that do not yet exist but which will come into being in order to tackle the formidable technological problems that lie ahead.

Though we are moving into a largely unknown area, one can get some notion of the impact of the expanding space program of the future on

---

[4] As of March 1961, companies with substantial NASA contracts included Aerojet-General Corp. (a subsidiary of the General Tire and Rubber Co.); Azusa of California; Ball Brothers Research Corp. of Boulder, Colorado; Bell Aerospace Corp. of Buffalo, N.Y.; Bendix Corp. of Detroit, Michigan; Brown Engineering Co. of Huntsville, Alabama; Chance Vought Aircraft, Inc. of Dallas, Texas; Chrysler Corp. of Detroit; Grumman Aircraft Engineering Corp. of Bethpage, N.Y.; Hayes Corp. of Birmingham, Alabama; Lockheed Aircraft Corp. of Burbank, California; McDonnell Aircraft Corp. of St. Louis, Missouri; North American Aviation, Inc. of Los Angeles, California; Radio Corporation of America, New York, N.Y.; Douglas Aircraft Co. of Santa Monica, California; General Dynamics Corp. of New York City; Thompson Ramo Woolridge, Inc. of Cleveland, Ohio; and Western Electric Co. Inc. of New York City.

industry by considering the past impact of the missile programs of the Defense Department upon industry and the economy. An analysis of missile production in fiscal year 1959 shows that the total outlay of $3.5 billion was split up as follows: 38 per cent for the assembly and testing of missiles; 34 per cent for electronics work, both airborne and ground; 16 per cent for propulsion systems; 5 per cent for non-electronic ground support—mainly stationary launching platforms; 4 per cent for nose cones; and 3 per cent for missile structure work.

Clearly, the percentage distribution of funds going for a future major space program will differ markedly from those indicated above for missile production in fiscal 1959. The necessity of shooting far greater weights far greater distances, and of keeping men alive in the hostile space and moon environment, will require tremendous outlays on a whole range of new things: the *Rover* nuclear motor, the *Saturn* series of motors, solid-fueled rockets, the *Nova,* new guidance and control systems, biological research, etc. It is obviously impossible at this point to predict the composition of outlays as compared with the pattern for the missile industry indicated above. However, the following table based on the guesses of space industry experts, may at least suggest the direction of change in components of the program through the 1960's:

PERCENTAGES OF TOTAL EXPENDITURE

	Missile Program	Space Program	
	FY 1959	1965	1970
Assembly & Test	38	25	15
Electronic ground support & airborne guidance and control	34	50	65
Propulsion	16	20	15
Non-electronic ground support	5	2	2
Nose cone	4	—	—
Structure	3	3	3

What the table is meant to emphasize is that a marked shift will be occurring toward heavier outlays on electronics work, with an initial shift (through 1965) on heavy R & D spending for propulsion, then a relative reduction by 1970; and a continuing decline in the proportion of expenditures going for assembly and testing—which might be accelerated if recoverable boosters or second-stages can be designed.

These shifts, particularly the heavily increasing share of outlays for electronics, are going to produce a battle royal in which the character of the aerospace industry will undergo some major changes and in which the position of the former aircraft companies will be under continuing challenge from electrical and electronics firms. Former aircraft firms aiming to survive in the space age have of course already been diversifying heavily into electronics. As D. C. Eaton puts it, "The airframe manufacturer can no longer lay any claim to any special position in the industry. Companies whose experience and growth has been in electronics are and will be equally, if not better, qualified to perform major system development."

In 1959, of the 16 companies that dominated the missile business and accounted for 72 per cent of the sales, eight were originally aircraft companies, six were electrical and electronics manufacturers, one was an automobile manufacturer, and one a subsidiary of a rubber producer. But the six top spots were all occupied by aircraft firms.[5] In 1961, of NASA's 18 principal contractors,[6] only eight were originally aircraft firms. Spectacular rises—and declines—of particular companies in the space business are likely to occur in the decade ahead. Consider what has happened within the traditional aircraft industry since the end of World War II. Curtiss-Wright Corp., the largest producer of aircraft in that war (measured in terms of pounds of airframe manufactured), had dropped to 53rd place in 1959; indeed, its production no longer included complete aircraft but was limited to engines and components. The Martin Co., which made a timely shift from aircraft to missiles, leaped from 23rd place on the list of defense contract holders in 1953 to eighth place in 1959. Defense is clearly a high-risk and unstable business; 21 of the 100 firms on the 1958 list of top defense contractors did not make the 1959 list.

Many industry observers foresee an intensifying battle for survival in the aerospace industry in the years ahead. Some believe that by 1970 the present number of close to fifty major companies in the aerospace industry will have shrunk perhaps to twenty or twenty-five; that the several hundred present medium-sized subcontractors will be markedly reduced as the prime contractors broaden their activities and pull in much present subsystem work; but that in 1970 there will still be thousands of small specialized companies providing specific and limited services at relatively low cost as there are today.

THE MANPOWER PROBLEM

Many people not close to the missile or space business have considered that space will be a low user of manpower. Nothing could be further from the truth. The *Mercury* project—costing in the neighborhood of $500 million by the end of 1961—has involved something like 4,000 companies which employed more than 200,000 people in the project. One cannot be sure how many manhours of labor were actually involved in *Mercury* without a detailed survey of all those firms, but it is safe to say that tens of thousands of full-time jobs were created.

---

[5] In order of rank, the top 16 missile companies were Convair, Boeing, Douglas, Martin, North American, Lockheed, General Electric, Aerojet-General (a division of General Tire), Raytheon, Hughes, General Motors, Western Electric, R.C.A., Northrop, Sperry Rand, and Arma.

[6] See footnote 4.

To obtain a rough idea of the employment effects of space, it may be noted that the somewhat similar missile industry is extremely labor-intensive. A study by the U.S. Department of Labor shows that during fiscal 1959, when the government was spending at a $3.5 billion annual rate on missiles, the number of persons directly employed in missile activity totaled 319,300.[7] This works out to a ratio of $10,893 in sales per worker in the missile industry, compared with $12,329 per worker in newspapers, $22,112 in construction equipment, and $99,566 in petroleum refining. The space industry is likely to be even more labor-intensive than missiles.

In the decade ahead space vehicles are not going to come off production lines with designs frozen for long runs over time; they are going to be jiggered and altered, almost item by item, both to incorporate continuously the results of scientific and technological progress and to design vehicles to accomplish different types of missions. All this implies a heavy use of manpower relative to the cost of the product. If the space program gets up to annual expenditure rates of $10 billion, and if the types of ratios indicated by recent experience hold, space will employ over a million workers directly—with important secondary employment effects on many other industries.[8] By comparison, automobile manufacturing in 1961 was directly employing about 800,000 and steel about 700,000.

Of course the manpower required for space programs will be in large measure highly specialized and skilled. The space programs appear certain to provide further momentum for the trend, so marked in recent years, for labor requirements to shift from low-skilled production workers to scientific, technical, and managerial occupations. The shift has resulted chiefly from the tremendous increase in expenditures for research and development. The R & D portion of the total cost of an intercontinental bomber was 20 per cent according to Weidenbaum's estimates; this share rose to 60 per cent for an intercontinental ballistic missile. For space systems the R & D proportion will go still higher; indeed, one might define space system work in the period that lies immediately ahead as virtually 100 per cent R & D. Companies which in the days of heavy aircraft production had two-thirds of their labor force made up of industrial workers will see that proportion

---

[7] U.S. Department of Labor, *Manpower in Missiles and Aircraft Production,* Industry Manpower Survey No. 93, August, 1959. This study defined missile activity as including research, development, and production of rockets and missiles, launching devices, ground control and testing units, propulsion units, warheads, fuel, electronic components, and other parts designed for inclusion in the missile or in ground support equipment. It included government-owned and operated research, development, and production facilities, but not those segments of the military services or NASA engaged purely in administrative activities such as planning and procurement. Nor did it include construction activities involved in the building of missile launching bases or test sites.

[8] A more careful estimate of the total employment effects of a space program of some given size and composition would require use of a detailed Leontief-type input-output model.

decline to perhaps one-fourth as the proportion of scientific, technical, and managerial personnel keeps rising. These shifts will create great opportunities for those people who can handle the new jobs and serious strains and dislocations for those who cannot. The shifts will impose a growing responsibility upon the American educational system to increase and upgrade the national supply of brains, talents, and skills.

Economic Benefits—and Risks

There can be little doubt that the most important element in economic growth is the expansion of scientific and technological knowledge which culminates in new products, new techniques of production or communication, new resources. This assertion is not a novel statement: John Rae developed the theme of the impact of new technology on economic growth very competently in the early part of the nineteenth century, and Joseph Schumpeter handled it even more brilliantly in the twentieth century. In our own time some economists have reached the conclusion that technological advance has accounted for about 90 per cent of the rise in productivity—output per manhour—in the United States since the latter part of the 19th century. The implications of these findings are of outstanding importance for policy makers and economists; they mean that the greatest emphasis in any program for long-term economic growth must focus on technological progress and the factors that promote it or obstruct it.

Technological progress resulting from space programs will stimulate economic growth basically in three ways:

(1) Innovations emerging from space programs will lead to increases in the productivity of both labor and capital and, consequently, to higher real income, either as a result of reducing prices or by permitting a rise in money incomes without an inflationary rise in prices. At the same time, from the standpoint of the firms that do the innovating, the savings in cost resulting from higher productivity will widen profit margins, making more funds available for expansion or research by the profit-making innovator, or will attract additional funds and resources from less profitable industries. Specifically, space programs will lead to the development of high-speed, lightweight computers which will replace the giant electronic brains of today; advances in "human engineering"; more efficient means of communication, through satellites; new and improved power sources—special types of batteries, smaller and safer nuclear motors, plasma power generated through the use of hot, ionized gas; cost-savings in many fields (such as agriculture and transportation) through better weather forecasting; better methods of preventing corrosion; better methods of adhesion; and progress in all areas of electronics which will have great relevance to advancing the efficiency of industrial production.

(2) By creating new products space research will stimulate demand, create new markets, necessitate additions to the stock of capital required to produce the new goods. Some space proponents have described as "space age" products such things as kitchenware made from ceramics that can stand extreme and sudden temperature changes; new battery-powered flashlights and radios that can be recharged by plugging them into a wall socket; remote control devices for opening garage doors, tuning radios or television sets, etc.; electronic wristwatches; heart stimulators, artificial larynxes, and other medical devices; monitoring systems for hospital patients which might both reduce hospital costs and save lives. Speedier and safer passenger travel may result from programs like the X-15 or *Dyna-Soar*, and from improved radar systems, engines, flight equipment, automatic pilots, improved weather forecasting. New "space" products also include such things as observation satellites or sensing devices, which increase the security and stability of the nation and the world and are fully as important to the firms that produce them as consumer goods.

(3) By making available new resources or by finding new uses for old resources space technology will increase the economic system's output and, through the price and profit mechanism, redirect resources to increase total production. Science and technology in our time have come to play the role that war and exploration formerly played in expanding a nation's economic resources. Space research is leading to the development of many new materials: synthetics and composites of synthetics and metals; new and very durable fabrics; reinforced plastics—silicones, polyesters; epoxy resins and phenolics reinforced with a variety of materials—asbestos, quartz fibers, graphite cloth, glass fiber, etc. Some observers believe that space research will underlie major breakthroughs in increasing the world's supply of traditional resources: for efficiently getting fresh water from sea water, for extracting minerals from sea water; for drilling more deeply into the earth for taconite (as high-grade iron ore becomes an increasingly scarce resource) or for other minerals; for replacing traditional energy resources with new fuels, solar energy, etc. Observation and sensing satellites and improved seismological instruments may provide a valuable additional means for discovering new resources on earth—or, conceivably, in outer space.

Obviously it may not be essential to send men on long-distance space missions to achieve many of these earthly by-products. And not all of the by-products can be ascribed solely to space research rather than to earlier or other types of research programs. Nevertheless, there can be little doubt that a major space research effort would accelerate this by-product "fall-out." As Hugh Dryden has put it, "Perhaps the greatest economic treasure is the advanced technology required for more and more difficult space missions. This new technology is advancing at a meteoric rate. Its benefits are spreading throughout our whole industrial and economic system." It is this great impetus which the space programs will give to scientific and technological advance in many fields—in electronics, metals, fuels, the life sciences, ceramics, machinery, plastics, instruments, textiles, thermals, cryogenics, and, most important, to basic research in all the sciences—that provides the reason one must consider the space effort not simply as an elaborate form of modern pyramid-building or pump-priming but, far more importantly, as a powerful ongoing force for innovation and economic growth.

The most striking economic development of the postwar period has been the way American businessmen and investors have caught on to the potency of research and development as a force for profits and growth; innovation—the celebrated elixir of capitalism—has become a widely recognized and increasingly diffused element in our system.

Competition among business firms to participate in the space program will therefore be intense. It will represent a real opportunity for companies to subsidize their own research programs. There are inherent excitement and challenge in the problems to be solved, and there will be prestige as well as profit accruing to the business firms that crack those problems. (After Commander Shepard's first successful *Mercury* suborbital flight, there was a great rash of newspaper and magazine advertising by companies that had had some share in the *Mercury* program.) There will be keen competition for the scientific talents that attract the government's funds. But, based as it is on the scientific imagination, space will be a rapidly-changing and high-risk business; the problems for business in this field are proportional to the tremendous opportunities ahead.

A Case Study: The Communications Satellite

Both the opportunities and the problems for business in space are well illustrated by the potentials of communications satellites (which are also discussed by Donald Michael in his chapter on Peaceful Uses). A wide variety of communications satellite systems is possible—passive satellites (like *Echo* I, the big mylar-skinned balloon, which just reflects signals back to earth); active (that is, satellites with electronic equipment to receive, amplify, and transmit messages); low-orbit; 24-hour high-orbit synchronous (satellites that, in orbit 22,300 miles from the earth, maintain a position over a fixed point on the earth); and others such as real time, delayed time, attitude controlled, and position controlled satellites.[9]

With so many possibilities of variation in the type of system to be used,

---

[9] In this discussion I am indebted to the RAND Corporation for permission to draw upon the study by William Meckling, "The Economic Potential of Communications Satellites," RAND P-2216, February 6, 1961.

the task of estimating future costs of satellite systems is obviously extremely difficult. It is further complicated by such additional variables as the probable useful life of various satellites, the probabilities of successful launchings, and the number of channels (for voice, television, or telegraphic signals) that can actually be gotten out of different types of equipment.

Even more difficult is the task of forecasting demand for international telephone, telegraph, television, and data processing services. The rate at which demand increases, is, in fact, the most crucial factor in determining how expensive or profitable international communications satellites will be. Undoubtedly the demand for long-distance communications will be increasing greatly over the years ahead; but the immediate question from a commercial standpoint is the *rate* of increase over *what* time period. The rate of growth in transoceanic traffic will be a function of the number of channels made available and of their cost; but it will also be a function of the rise in the volume of international trade and investment (both direct and portfolio), the integration of business operations of companies operating in different parts of the world, international cultural relations, rates of national economic growth, diplomatic agreements, and so on.

The communications satellites will provide enormous increases in capacity for international communications. In 1961 the number of telephone circuits to Britain numbered 57, to France 21, to Germany 19, to Italy 8, to Central and South America 40, to Caribbean points 147, to Hawaii 43, to Japan 8, to Australia 2. By comparison, a world-wide low-altitude satellite system with 7800 voice channels would provide about 2200 per cent more international channels than those listed above, and a single 24-hour satellite with 4800 voice channels would provide almost 1400 per cent more. The increase in international telephone traffic to and from the United States in the last ten years was only 300 per cent; international traffic today is estimated to be increasing by about 20 per cent per year.

Therefore Meckling considers that *estimates of* satellite communications costs, calculated on the basis of full capacity operations, appear to be "highly optimistic," since it may be years before satellites are used to full capacity—unless, of course, substantial rate reductions are affected and provide a very great stimulus to the increased use of international communications systems by individuals, businesses, and government organizations (and also if military communications make substantial use of commercial satellite facilities).

On the basis of full capacity use, the RAND study finds that communications satellites will make for great cost savings. For a low-altitude communication satellite system, when the mean time to failure of a satellite is assumed to be two years and the launch success probability is assumed to be .75, a price of about $8500 per year per channel would cover all costs. Similarly, for a high-altitude synchronous system, when the mean time to failure is taken as one year and the launch success probability is estimated at .75, a price of $10,000 per year per channel would also cover all costs, assuming that all the available channels were used. By comparison, for new submarine cables the annual cost of a 2500-mile transoceanic link is about $24,000 per circuit. Telegraph companies pay $240,000 per year for one voice channel in the existing submarine cable.

Thus the potential savings through satellites are tremendous should demand expand adequately. Even with relatively conservative assumptions on the buildup of international communications traffic the RAND study finds that the use of communications satellites should be competitive with submarine cables over the next 15 years. Assuming that, with a low-altitude system, 1,000 channels are sold in the first year of operation and that thereafter sales increase at the rate of 15 per cent per year, reaching capacity operation (7,800 channels) in the 15th year, Meckling calculates that the annual cost per circuit of the low-altitude system would be $24,000 per year for a 2,500 mile transoceanic link, or the same as a submarine cable would presently cost. The cost of a 24-hour system would be slightly higher, assuming that each synchronous satellite stays operational for only one year. If, however, a more optimistic assumption should be justified on the mean time to failure of a 24-hour satellite system, its cost per channel would also be fully competitive with submarine links. R.C.A. presently prefers the synchronous system; A.T. & T. prefers the lower-altitude random satellite system.[10]

Calculations by private industry have produced similar results to those reached in the RAND study; several such studies have shown that despite the high initial costs of a communications satellite system—estimated at $400 million or more—a satellite system could become profitable before the end of the decade, and, within 15 years, could become a major contributor to the sales and profits of even huge corporations.

Communications satellites are of great importance, both because they can expand the number of channels for international traffic enormously, quickly, and economically and because they will permit types of international communications, such as television and massive data processing, which would otherwise be technically unfeasible. The political as well as the economic consequences of this stimulus to international integration and development are bound to be of very great economic and political significance. It is no wonder that there has been such intense competition among both communications companies and equipment makers to win contracts such as Project *Relay* and thereby to gain a foothold in space communications. Many companies have seen the importance of entering major space fields in the early stages of their development in order to get funds for the high-risk research involved, to keep up fully with technological advances,

---

[10] *The New York Times,* June 4, 1961, p. 52.

to be in a position in the future to meet government specifications, and so to be prepared to move ahead with commercial programs.[11]

Considering the possible effects upon the strength and influence of the companies concerned of allocation of the initial contracts for space R & D, the government must seek to devise new structures that will permit the orderly advance of research programs without prematurely closing the door to organizations attempting to enter. Some companies are determinedly moving ahead with space communications programs of their own, unwilling to trust the outcome of space competition to the chances of government contract-letting.

Communications satellite systems are going to be extremely expensive; some consider that their costs may be beyond the resources of any single company. Several companies and groups of companies have made proposals for sharing arrangements. Government itself has of course a major stake in any system that is devised—not only because it has already paid for so much of the underlying R & D involved but also because it has heavy space communications requirements (particularly for the military services) and must in any case take responsibility for launching and monitoring or controlling satellites.

Setting up space communications systems will pose complex institutional, regulatory, economic, and operating problems similar to those involved in the fields of nuclear and electrical power—such problems that already have generated vast quantities of heat relative to light, as during the long-continued controversies over the Tennessee Valley Authority and in the general debate over public versus private power.

But a new ingredient has been added. For communications satellites are inherently international in character. This means that the resource planner seeking to optimize the use of radio frequency resources must give equal consideration to the complexities of working out relations with the businesses and governments of other nations. (The same will be true for the use of weather or mapping or exploration satellites as well as for re-entry transport, patterned after *Dyna-Soar*.) The complex negotiations necessary for setting up *Mercury* tracking stations give only a minor indication of the international negotiating problems that lie ahead for space communications and other space programs. It is quite possible that the international administrative, legal, and operating problems may be even more difficult to settle than the still considerable engineering problems involved in exploiting space commercially.

Very probably the international regulatory and diplomatic problems will slow the commercial use of space. It may be a long time before the Federal

---

[11] I wish to acknowledge my indebtedness to Mr. James Williams of United Research, Inc. and to Mr. Jack Oppenheimer of NASA for their thoughtful discussions with me of the political, commercial, and other complexities of building and operating an international communications system.

Communications Commission and its foreign counterparts work out the forms of international organization and operation needed for communications satellites; there are difficult domestic as well as international problems involved in the use and assignment of frequencies. Businessmen will have to be cautious and probably conservative in projecting the dates when the machinery of government and diplomacy will be oiled to permit a space system to be used. This means that individual companies will have to be chary of freezing operational designs too soon lest "the state of the art" may pass them by while the administrative machinery grinds away. Bad timing might enable a competitor to overtake and pass a company with several years of prior investment and an apparently safe lead.

The international character of space communications must constantly be uppermost in American business planning in this area. The American company that does not conceive, design, and build systems that can be integrated into existing facilities in other countries will find a narrow market, one that will be far short of expectations. Catastrophic losses may hit companies that fail to take full account of the international nature of space systems and the individual country differences affecting system design. And not only domestic but also foreign competition for the building and operation of space communications systems is bound to be intense.

Communications satellite systems thus typify the problems, risks, and uncertainties facing business organizations in outer space. Yet the problems should not—and will not—obscure the vast economic opportunities that lie ahead in space. Satellite communications systems are only one dramatic instance of why a growing number of observers are coming to regard man's venture into space as the great enterprise—economic as well as political and scientific—of this century.

APPENDIX A

SPACE ACTIVITIES OF THE UNITED STATES GOVERNMENT
New Obligational Authority/Program Basis—in millions
Historical Summary and FY 1962 Recommendation and Congressional Adjustments

	NASA [1]	Defense [2]	AEC [3]	NSF [4]	WB [5]	Total
FY 1955	$ 56.9	$ 3.0	$ —	$ —	$ —	$ 59.9
FY 1956	72.7	30.3	7.0	7.3	—	117.3
FY 1957	78.2	71.0	21.3	8.4	—	178.9
FY 1958	117.3	205.6	21.3	3.3	—	347.5
FY 1959	338.9	489.5	34.3	—	—	862.7
FY 1960	523.6	560.9	43.3	.1	—	1,127.9
FY 1961	964.0	751.7	62.7	.6	—	1,779.0
1962 Budget, 1/16/61 ..	1,109.6	846.9	55.1	1.6	2.2	2,015.4
Increases recommended 3/28/61	125.7	159.0	23.5	—	—	308.2
Increases recommended 5/25/61	549.0	77.0	7.0	—	53.0	686.0
Total FY 1962 recommendation	1,784.3	1,082.9	85.6	1.6	55.2	3,009.6
Specific Congressional adjustments	—112.6	85.8	—	—	—5.0	—31.8
FY 1962 as approved by Congress	1,671.7	1,168.7	85.6	1.6	50.2	2,977.8

[1] National Aeronautics and Space Administration amounts are totals for all activities of NASA and include totals for NACA prior to establishment of NASA.

[2] Department of Defense amounts are based on identifiable Defense funding for space and space-related effort and do *not* include substantial amounts for (1) construction and operation of the national missile ranges with regard to space programs, (2) the cost of developing missiles such as *Thor* and *Atlas* which are also used in space programs, or (3) supporting research and development (such as bio-medical research) which is more or less mutually applicable to programs other than "space."

[3] Atomic Energy Commission amounts are those identifiable with ROVER nuclear rocket and SNAP atomic power source projects.

[4] National Science Foundation amounts are those identifiable with VAN-GUARD and with the NSF space telescope project.

[5] Weather Bureau amounts are those identifiable with the metorological satellite program.

Bureau of Budget, October 30, 1961

4.

The task for government

♦ T. Keith Glennan

This chapter will examine some of the major problems that were faced by the United States in undertaking the initiation and management of the technology necessary to achieve a position of pre-eminence in the exploration of outer space. To a considerable extent the author has found himself applying the experience had in this particular activity to the problems that will be encountered in the undertaking by the government of extensive activities in any major new field of technology. There are certain dangers inherent in attempting to draw general conclusions in this area from a too complete reliance on the experience in the space field so far by the National Aeronautics and Space Administration. Conditions attending the birth of this organization may or may not be repeated. The glamorous nature of the venture, the semi-hysterical reaction on the part of many organizations and people to the initial successes of the Soviet Union and the subsequent clamor for action (almost any action), the prior and understandable interest of the military services in this particular activity, the availability of prior art and devices in the field of rocketry in this country—all of these combined to give a somewhat unusual setting for this national effort.

Yet one can draw from the NASA experience some conclusions and pose even more questions which will highlight the major problems that have been faced by this new organization. Indeed, some of the questions posed will continue to be the concern of the management of NASA and of other agencies of the government for some time to come. Since this paper is intended to stimulate discussion of the many issues involved, a wide range of subjects will be scanned and a number of the questions will be left unanswered; for at this stage of the nation's program of space exploration it seems more worthwhile to point up the problem areas than to state definite conclusions concerning them. Concrete examples of the problems faced will be cited from NASA's experience. In so doing there is intended no defense of the actions taken or criticism of the policy decisions made even though, as a major participant in these activities, the author properly may be accused of some personal bias.

It should be borne in mind that extremely high stakes may be involved in whether or not a major new technology is effectively organized and managed. In the last two decades the world has seen the dramatic appearance of two new technologies that illustrate the great significance of new technology in terms of world leadership. Failure to gain, or to maintain, the lead in the use of a new technology can severely impair the world position of the United States or the Soviet Union. Success, on the other hand, almost overnight can alter the balance of world leadership and influence, at least for a time.

The importance of new technology to the United States is heightened by the fact that we have in the Soviet Union a competitor with demonstrated skill in using new technology to its own advantage. The Soviet leaders have made it abundantly clear that they intend to outdo us to convince the world that their technological superiority is another significant indication of the inability of our political and economic systems to meet the needs and to rise to the challenge of the world of the future.[1]

Defining the Goals

The establishment of goals should be a pre-condition to the initiation of any major project. In one which by its nature and cost is peculiarly the responsibility of the federal government it is mandatory that the goals be defined if the nation is to understand the program and to give its full support. In the nation's space exploration program the pressures for action made this a difficult task. Indeed, it is clear at this writing that much remains to be done in charting the course in space for NASA and the nation. A so-called *Ten Year Program* was first defined and reported to the Congress in January of 1960. That program was revised and up-dated in January of 1961 and will require periodic revision as the years roll on. But the big decisions—those that may require the commitment of several billions of dollars each year for many years—have not yet been made.[2]

To an extent, this is natural. Only now are we beginning to understand the magnitude of the task of exploring outer space with both manned and unmanned space craft. Only now is it apparent that for many of the most difficult and exotic of the missions a considerable number of years of hard and costly effort must elapse before the missions can be flown. For most of these tasks the basic technology is at hand, although much design and testing remains to be done. It is to be remembered that the bulk of the money and effort to be expended will be devoted to straightforward systems and engineering design—not scientific research. What we are doing is developing very expensive "test tubes" with which the scientists in many fields can pursue their researches. They—the scientists—set the tasks, and the engineers must provide the means for accomplishing them.

It is possible to pose some specific questions. Should our energies and funds be devoted to the accomplishment of a few spectacular shots such as landing a man on the moon at the earliest possible date? Acknowledging that this objective one day will be achieved, is it in the nation's best interests that the program be concentrated principally on this objective? Or is it possible to recognize that to accomplish this single objective we must learn a great deal more about outer space—the hazards that must be understood and guarded against, the amount of shielding needed to protect the astro-

[1] The author acknowledges the considerable assistance of a former colleague at NASA, Mr. Walter Sohier, in the preparation of this discussion.

[2] At the time of writing it was too early to judge public and Congressional reactions to President Kennedy's proposals to the Congress implying that the United States was entering a race to be first to the moon.

naut from the deadly radiation concentrations through which he must pass, the problems of landing and subsequent take-off for the return journey, and so on? Under present circumstances can we afford to pursue an orderly course of accumulating knowledge, bit by bit, that will permit us to undertake at some future time, with reasonable assurance of sucess, the longer-range task of landing men on the moon? Are not the shorter-range objectives of developing useful applications of space technology in the fields of communications, meteorology, and navigation—activities that promise real benefits to mankind—equally or even more important as national objectives?

Much has been said and written by partisans of both points of view. Always there lurks in the background the spectre of Soviet competition. And since the main thrust of Soviet efforts to date seems to be toward the accomplishment of manned flight into and through space, there is a very vocal clamor to "beat the Russians to it." There is a hard decision to be made here. The easy answer is to order an all-out effort to land a man on the moon. But it is not clear that for our nation that is the best or even the right answer.

To make really intelligent policy decisions in this area of government activity there is needed a basic guide to the national interest in space. If we had such a guide—a set of national goals—we would be able to decide on the relative priorities among conflicting claims for highly skilled manpower and for financial support. As it is, we tend to answer the challenges of others rather than to set a course that would be consistent with what should be the major elements of national policy.

Perhaps the problem can be stated in another way. If our competitor does not play the game as well as we do or is not determined to win, we can play at a leisurely and haphazard pace without any particular guiding strategy. But if he is our equal and is desperately anxious to win, and if the stakes may be the survival of the principles in which we believe, the game takes on a different complexion. Then we must decide whether we shall try to match him on his grounds—shot for shot—or pursue a course which will provide the best chance to demonstrate to ourselves and the rest of the world the superiority of the open and competitive society which embraces those principles. To pursue the former course is to run the risk of consistently running a stern chase since the competitor's objectives are not announced and are revealed only upon completion of a particular phase of his effort. To pursue the latter requires convictions about our national goals supported by courage and leadership of a high order. Which shall it be? Or can we and should we set out on both courses?

These and other questions are troubling many of our responsible citizens today. Can the relative success of the United States in meeting past challenges imposed on it be expected to continue in the fast-moving world of modern science and technology? Can we devise some change in our pattern of relying upon response to crisis? How can the United States act to meet the challenges of modern technology effectively and retain its technological leadership without endangering its free institutions? Such questions have been given new urgency by the advent of space technology.

It would be presumptuous to try in this chapter to answer these questions, but we can say that, as a minimum, a major educational effort and a broadly based program of science and technology are needed to enable this country to respond rapidly and decisively to emergency. Such a broad attack would help assure that the United States has enough highly motivated people with the necessary skills to meet the bold new opportunities presented by science and technology. NASA has set about this task so far as education in space technology is concerned, but its role must be limited to supplying information and guidance and stimulating interest and activity. It may well be that what is required is more of a crusading spirit on a national scale in education and in the fields of science and technology if we are to meet the challenge posed for us by the Soviet Union. Moved by such a spirit, we may be better able to give realistic expression of our national objectives.

Organizing for Management

In considering the problems of organizing to manage the nation's program in space exploration three general subjects will be discussed: (1) assessing the type of management job involved, and locating and utilizing the people and resources required to carry out this job; (2) selection of the type of organization to manage the program; and (3) defining the specific military objectives and responsibilities. All three are interrelated. Both the type of management job involved and the extent of military interest in the development and use of the technology will affect the choice of management organization.

ASSESSING THE MANAGEMENT JOB

It may sound elementary to mention here the importance of making an early assessment of the management job involved in exploiting a new technology. Yet in the space program, if there were an opportunity to turn back the calendar three years, greater attention would probably be paid to this task than was paid during the formative period of NASA. It is evident that different technologies will pose substantially different types of management jobs and problems. Moreover, it may be difficult at the outset to determine just what the management job is and will become in the years ahead.

The management job that has evolved for NASA is that of directing a substantial development program performed largely under contract with industry. The skills needed to manage such a program are different from those required in the running of an essentially in-house research program of the type conducted by NACA—the old National Advisory Committee for Aeronautics—during its life as an organization in its own right. As NASA's program has progressed, the full thrust of this need has become evident, necessitating the modification of some organizational decisions that were made initially. There has resulted a need for skills in project management and systems engineering that are, or were, in short supply in the organization.

Another example of the difficulty of assessing the job at hand relates to the educational and technical information programs of NASA. The magnitude and importance of these functions have become increasingly evident as the space program has evolved. For example, the area of collecting and disseminating technical data is a vital concern.

Actually, the principal product of NASA may well be the new information derived from space research and the development of gadgetry and sophisticated devices to accomplish these researches. The program is essentially unclassified, and the legislation requires the widest practicable dissemination of information. Recipient groups, such as the general public, the scientific community, and industry, require different treatments of the same or similar information. It must be distributed in a timely manner if it is to be most useful in furthering the development of new devices and avoiding the gross duplications of effort that might otherwise occur.

The space program requires for its support broad and enlightened public understanding. Government agencies may not use public funds to lobby for additional support. But their information offices have a major responsibility to improve public understanding of their program. Further, in a new area of research and development, school children and their teachers should be encouraged to understand the new technology, its meaning and ultimate importance. Hence, there is required the preparation of pamphlets, display kits, lecture and classrom notes, etc.

Undoubtedly an earlier start would have been beneficial. Viewed several years hence, even today's active program may appear to be a cautious beginning.

SELECTING THE MANAGING ORGANIZATION

Deciding the type of organization that should manage the nation's effort to exploit a new technology must also be resolved at the outset. In reaching this decision the extent to which federal government participation is required must be determined. The very size and nature of the nuclear energy and space programs made it easy to decide the extent of federal participation in management. It seems likely that any future national effort on a similar scale will also require, at the outset at least, management or substantial control by some agency of the government.

The type of government organization to select for manager will require a legislative decision either in the appropriation process or through legislation establishing a new agency and giving it the necessary authority and money. Moreover, the national aims to be achieved by the new technology will affect the choice of managerial organization. This is particularly so in weighing the alternatives of creating a new organization or utilizing an existing military capability.

In essence, there are four alternative choices for selecting the type of governmental organization. *First* is the possibility of using an existing organization, such as the Department of Defense or a single military department. Such a choice was carefully weighed at the outset of the space program. *Second,* an entirely new organization might be formed within the government. The Atomic Energy Commission is one example. *Third,* facilities and personnel already existing within the government might be utilized as the foundation of an essentially new organization. This route was adopted for the space program. *Fourth,* the government might be limited to performing a regulatory, or perhaps only a promotional, function, leaving responsibility for exploitation of the technology in non-governmental hands with or without governmental financial support.

There were several reasons for selecting the third alternative for the space program. There was in being the National Advisory Committee for Aeronautics, with first-rate engineering and some considerable scientific talent as well as excellent facilities at four research centers in the United States. NACA had already been engaged in work involving space science and technology as a natural outgrowth of its work in aeronautics. It enjoyed a high reputation in industry and government circles. To utilize it for the space program would save time, a valuable factor in view of the atmosphere of urgency in which the space program had been launched. Moreover, future use of the existing facilities and personnel of NACA for aeronautical research was on the wane as the missile era progressed and new programs for military manned aircraft declined. These personnel and facilities seemed to fit logically into the new horizons of space, and a valuable national asset could thereby be preserved and reoriented.

Furthermore, NACA's history of effective relationships with the military departments and other civilian agencies of the government, as well as the scientific and technical community generally, seemed to make it a natural choice. It was felt that this reservoir of good will would prove valuable in getting maximum cooperation and assistance from other agencies of government, particularly the military.

These advantages have, in general, proved themselves out in practice. For example, existing business staffs inherited from NACA were there to handle the new and expanded contracting activity of the new agency. These business staffs needed supplementing with personnel experienced in this type of program, but contract action was able to begin without delay both in Washington and at the former NACA field installations. Technical programs could be planned and carried out utilizing the talents inherited from NACA, again saving valuable time in initiating such projects as development of the F-1 engine and fabrication of the *Mercury* space capsule. The transfer from military laboratories to NASA of several programs started by the military was accompanied by the transfer of capable personnel, thus further strengthening the new agency.

The reasoning behind the organizational decisions made for the space program has been set forth in detail to illustrate the types of considerations, not to justify the decision.

DEFINING MILITARY OBJECTIVES AND RESPONSIBILITIES

Since the over-all top-level direction and control of the nation's defense program are in civilian hands, it may be an oversimplification to refer here to military and non-military interests, particularly in the present cold-war climate when all our resources and interests are involved in an across-the-board struggle. Nevertheless, this distinction has been preserved to point up certain important issues.

New technology has such important implications for the defense posture of the United States that it is difficult to conceive of a major new technology in which some military interest would not exist. There may, however, be serious drawbacks to permitting new technology to be exploited under military auspices. Accordingly, it is important that the military interest, and the way in which it is to be injected into a program, be faced up to during the initial stages of organizing for management. This too is a continuing problem requiring constant coordination and the making of timely decisions as to military and non-military responsibilities if the confusion which can so rapidly develop in this area is to be avoided.

Thus far the space program has had relative success in dividing responsibility and activity between NASA and the Department of Defense. First of all, in the initial legislation establishing NASA, Congress faced up to this problem and set the policy that the exploration of outer space was to be undertaken as a civilian activity for peaceful purposes for the benefit of all mankind. Accepting the fact that this newly available environment held potential values for the military defense of the nation which would most properly be exploited by the Department of Defense, Congress then provided machinery to resolve disputes through the mechanism of a Space Council and a Civilian-Military Liaison Committee. In general this machinery was effective during the initial period. For several years after, it was not particularly active, probably because management was proceeding reasonably satisfactorily. Most recently it has made recommendations on the questions of public policy involved in making operational a communication satellite system.

Timely decision-making has also been effective in resolving questions of jurisdiction before they could lead to difficulty. Some legislative improvements were proposed by NASA in 1960, in light of its experience and the development of its program, to make clear the responsibility of NASA for initiating and carrying out the nation's program of space exploration and to simplify the organizational structure. These proposals were passed by the House of Representatives, but the Senate failed to act on them. Recent changes in the Space Act of 1958 relating principally to the character and composition of the Space Council have not yet been tested in practice.

In spite of such relative success in sorting out the military and non-military projects, it is easy to see the unfortunate consequences that can result for the space program, just as for any other future new technology, if the government is not alert to this problem. First, wasteful, as distinguished from conscious and useful, duplication of programs can result. The F-1 engine is a case in point. If responsibility for this large-thrust engine had not been placed on one agency, unnecessary and duplicating programs might have resulted. The same is true with communications and meteorological satellites, where practical decisions dividing responsibility for programs were made at the outset. These satellite programs also illustrate the need for continual review of such decisions. Such a review was undertaken at the time of preparation of this paper with respect to communications satellites in view of their potentialities for commercial application. Still another example of avoiding duplication was the decision of NASA to terminate development of its *Vega* launch vehicle in favor of the *Agena-B* development program of the Air Force.

A second unfortunate consequence is the public confusion that may result concerning where responsibility lies. This in turn can lead to a lack of public confidence in the program, stemming from a belief that wasteful duplication exists, and to a failure of adequate support.

Finally, lack of a clear line between military and non-military interests may have serious international implications.

NASA Management and Operating Problems

Our discussion of management and operating problems faced by NASA falls under five general headings: determining the portion of our national resources to apply, and the relative priority of elements within the program; special industry problems; problems stemming from public opinion at home and abroad; the difficulties of hiring and retaining the people required; and international implications.

ALLOCATION OF NATIONAL RESOURCES

Allocating national resources can be considered at two levels: first, the share that exploitation of this new technology should be given as a claimant for a portion of the over-all national resources; second, deciding how to allocate the funds made available to various elements of the program. The ensuing discussion does not attempt to make a clean separation between these two aspects.

Determining the appropriate portion of our national resources is closely related to the matters already discussed. Does this allocation of resources occur more as a result of a national response to a crisis or challenge imposed upon the United States than from a deliberate plan based on the national objectives? Is a system that determines what portion of our resources to devote to new technologies only in response to sudden need of crisis adequate to meet the challenge that the Soviet Union poses in new technology? Is any other system of allocation possible under our system of government and type of economy? These fundamental questions are not to be discounted.

Aside from these general considerations, space exploration does give rise to some difficult practical problems in terms of commanding adequate support for effective management. For instance, it may be characteristic of a new technology that it is difficult to point to concrete benefits that will accrue from the program at any particular stage in its development. This factor is present in the space program in an interesting way. With respect to the main purpose of the program, exploration for the benefit of mankind per se, it may be difficult to demonstrate concrete benefits likely to accrue. Such benefits tend to show up more clearly in connection with programs that are a little to one side of the main street of NASA's space effort. For example, concrete benefits can safely be predicted for the applications programs, which involve the meteorological and communications satellites. It is also evident there will be significant side effects of space technology having considerable value for our economy. Examples of such side effects have already appeared, such as in the development of high-speed, light-weight computers.[3] The potential difficulty is that of keeping a program on course to achieve its ultimate objectives if concrete pay-offs are hard to demonstrate at a particular point in time, or if some side aspect presents the only tangible prospect of commercial or other reward.

---

[3] This whole matter of practical benefits is fully explored in a report of the House Committee on Science and Astronautics: "The Practical Values of Space Exploration" (House Report No. 2091, 86th Congress, 2nd Session).

The problem is common to most areas of fundamental scientific inquiry —the general public is conditioned to the expectation of results and does not understand that money spent for research is "investment" money, not an expense item.

A second problem, which arises in part from the way that the United States normally gets started in exploiting new technologies, is the difficulty of obtaining the optimum portion of our national resources to maintain a soundly based program. For example, at the present time, with public opinion aroused to beat the Russians, any difficulty in establishing tangible benefits to be derived from much of the space program is largely offset by a national desire to make progress in a space race. The elements of national prestige are so obviously at stake that the current pressures are to increase our space effort. However, if a Korean-type of emergency should suddenly arise, how much public support would there be for continuing with an adequate program for the exploration of space in light of such a large new drain on taxpayers? How much of our national resources would then be determined as appropriate to devote to space activities?

The danger here is that of feast-or-famine financing. NASA has seen such a tendency. At the outset it has been generally showered with kindness. The pressure has been to spend more and to request more. NASA's concern has been over whether such interest is only skin-deep. Its broadly based program that avoids a race psychology, without overlooking the value of occasional "firsts," has been drawn up with the long pull in mind. And NASA must educate the public to accept such an approach to exploiting space technology if a sustained program is to be followed. Such a program has in fact received considerably more emphasis within NASA in the past year although the first success in the *Mercury-Redstone* manned flights changed this somewhat. NASA has learned from experience the importance of instituting at the outset a fast, broad, and full program of education via all media. Close liaison with educational institutions is also required in this effort.

A third difficulty in the allocation of national resources is the almost impossible task of assessing with any degree of accuracy the amount of resources needed to carry out program objectives. A new technology necessarily will require reliance on cost-type contracting, which by its very nature makes budget planning difficult. Moreover, where research and development are involved, there is an inherent uncertainty as to the rate at which money can effectively be spent. Sudden new avenues of approach may open up, requiring substantial funds not originally anticipated. NASA is faced with these problems.

Finally, such a new technology must decide how to spend the resources allocated to it. This is closely related to the problem of obtaining adequate understanding and support from the public and the Congress, since certain phases of the program may lend themselves to more popular support than others regardless of the end objectives. It is an indispensable prerequisite to making a sensible allocation of funds to have worked out a reasonably clear concept of the end objectives and the real purpose of exploiting the technology. For example, NASA had to decide such questions as the following: What are the most important things to do first in space? Is putting a man into space via Project *Mercury* more important than other programs not involving a man in space? It is difficult to decide these kinds of questions intelligently if ultimate objectives are obscure. In the NASA program the fence was straddled, and both the manned flight program and broadly based research were instituted. The *Mercury* manned flight program was one of three given the highest national priority for materials and production facilities. The other two were the *Vanguard* and *Saturn*.

SPECIAL INDUSTRY PROBLEMS

The discussion of special problems bearing on government-industry relationships in managing a major new technology will touch on four separate subjects: the need for giving industry clear statements of policy and guidance; the issue of contracting out versus performing work in-house; the area of commercial applications; and the field of patents.

New technology poses both opportunities and problems for industry. The forces at work in our economy may suggest that industry become involved early in the technology. Yet in most cases the management of the program is in the hands of the government, as is much of the information concerning the technology. It is imperative therefore that the government make clear policy statements to industry at the outset and concern itself both with avoiding the misleading of industry and with making available to industry adequate information for it to make intelligent decisions. Otherwise industry may become overcommitted at too early a date in both its own and the government's programs in terms of personnel and facilities, or in work that will not ultimately justify the extent of the commitment. Although such decisions are industry's to make in the last analysis, the government has a responsibility to provide useful information and clearly state its plans to the extent it can. The present situation in the exploitation of peaceful uses of atomic energy is a case in point.

Some examples from the space program might be useful to mention. To head off any danger of overcommitment by industry of its own resources at this point in time, public statements of NASA have emphasized the particular characteristics of the space program that might create such a problem for industry. For example, the prospects of any substantial production follow-on stemming from NASA's development programs are not readily apparent at the present time. This points up an important difference for industry from the military programs with which it has been involved, since the production phase is most often the phase where a company realizes its earnings. Other new technologies of the future may present similar or related types of problems. Government must explain these special problems clearly to industry both at an early stage and throughout the life of the program, answering as early and as best it can such questions as: Will it lead to large production orders? Or is it essentially an R. & D. program? Should industry invest large sums in new plant, environmental test facilities, retraining men, and so on?

A second example, relating to a somewhat different aspect of government-industry relationships, stresses the importance of making clear statements of policy to industry to reduce confusion and apprehension. When NASA was formed it was feared that, with a new agency entering into a substantial procurement program, industry would be required to learn new rules in order to do business with this agency. Furthermore, the industry involved was largely the same industry with which the military departments had been placing their own development and production contracts. Within days of its formation NASA issued a reassuring statement to industry that NASA would adopt substantially the same procurement policies and procedures as those of the Department of Defense. Subsequently NASA made clear its intent not to have duplicating staffs to administer its contracts but, instead, to utilize the services of the military departments for this purpose.

A second major area of interest, and potential friction, for government and industry is the extent of contracting-out-to-industry contemplated as compared with the work to be performed in-house by the government agency itself. Should all design work on *Mercury* be done in governmental establishments, or should it be done by industry under general specifications prepared by NASA? Should basic research be concentrated in governmental laboratories or should the universities be encouraged to join in this effort—and on what basis? Since this decision has evident internal organization implications for the managing agency, it behooves the agency to make the decision at an early stage.

A government agency cannot contract out the entire management of a new technology. It must have the competence to specify the tasks it wants performed by industry. It must be able to monitor both technically and from a business standpoint a contractor's performance, and to evaluate the degree to which a contractor has met specifications in the delivered product. To do this may require the retention of selected portions of the work in-house to assure that an adequate level of technical competence is available within the government. Early in its history NASA had to face the problem of determining the extent and character of the in-house work it should retain to be able to manage its program effectively.

Another factor affecting the extent of "contracting out" is the relative capability of industry and government in personnel, facilities, and experience to do the job at hand. One objective, of course, is to utilize available national resources in the most effective manner whether these exist in industry or within the government. For example, the NASA research and space flight centers contain important national facilities for the conduct of space research. Its people are skilled and dedicated. What would be the best method of utilizing this capability in conjunction with the rest of the NASA program and the resources of industry? Should these facilities be operated for NASA under contract with industry or the universities? Should the level of effort inherited from NACA at these centers be continued without any expansion, with the balance of NASA's requirements to be obtained by contract with industry? Or should the personnel and facilities be substantially expanded in preference to reliance on industry? NASA's decision to retain the current level of effort at the centers was based on essentially practical grounds of availability and economy and the desire to avoid adding large numbers of people to the government payroll.

Third, the key question to be resolved in the area of possible commercial applications is the role that the government is to play. Is the government or is industry to finance the research and development effort? Should facilities be made available to industry by the government, and, if so, on what conditions? What regulatory or licensing requirements must be imposed by the government upon such commercial activity once it becomes operational? What intra-governmental relationships are most effective where two or more agenices have direct interests in different aspects of the new application?

The best way to point up these questions is to take a case in point—the communications satellite. American companies have indicated a definite and immediate interest in such satellites to provide world-wide commercial communications. At the present time both the Department of Defense and NASA have communications satellite projects. The immediate goal of the research and development program of NASA is to demonstrate the technical feasibility of using communications satellites as integral elements of a long-distance communications system. Such developments, however, must fit into the balance of the communications system, which, as an operating facility, is owned by commercial companies of the United States or by foreign governments or foreign commercial concerns.

Unlike the normal situation in this country where communications improvements are privately financed, in these satellite projects the government is directly involved in the research and development of the spacecraft and related facilities and components that eventually may well be utilized by industry as part of its operating system. How far along the research and development cycle should government financing continue? Can industry take on such a program if the government does not defray part or all of such costs? If the cost of research and development is shifted to industry, what are the national and international ramifications? For example, do the international political implications require that the feasibility of communications satellites be first demonstrated under governmental auspices? How important is it to achieve an operational communications satellite at the earliest possible time? Would commercial financing of the remainder of the research and development effort jeopardize the time schedule in such an event? What problems arise from the possibility of creating a monopolistic situation, and what role should the government play in view of such a possibility?

These questions exemplify the issues that must be resolved concerning the research and development of one type of satellite. When one ponders the operating phase, a host of other questions are presented. Assuming that the ground terminal facilities and the spacecraft would be paid for and owned by the commercial concerns, what about launch facilities and launch vehicles? Would the government require that launchings take place from its own sites and under its supervision, or would separate commercial launch facilities (at Cape Canaveral, the Pacific Missile Range, or elsewhere) be permitted? If the latter, what government regulatory measures would be necessary? What are the problems posed by permitting use of government launch facilities in terms of charging fees, scheduling, and the like?

Quite a few government agencies have an interest in this matter. There will be a series of problems in communications regulations involving the Federal Communications Commission. Other regulatory functions (launch schedules, for instance) will be involved that might ultimately be made the responsibility of NASA. The Federal Aviation Agency will be concerned since the launchings will be through the air space. As indicated earlier, the Department of Defense has projects in the field of communications satellites. Finally, the Departments of State and Commerce will be involved in one way or another.

It is in the national interest that the issues regarding the government's role be resolved rapidly. Otherwise confusion and a waste of public and private resources can result. This presents another illustration of the need for providing industry with clear statements of policy and intent. However, it may be difficult to think such matters out before knowing a good deal about them from a technological standpoint.

Fourth, the Space Act brought to the surface again basic differences of opinion among elements of the government, the Patent Bar, and industry. These were voiced at the time of enactment of the Atomic Energy Act and, in the following year, in the Attorney General's report concerning the rights that should be retained by the government as to inventions made under government contracts. Underlying a determination of the rights that the government should acquire to inventions made with government financing, or under government contracts, is the question of what the government really needs in order to be able to exploit the new technology involved. Does the government need to acquire title to such inventions, or is an irrevocable, nonexclusive, nontransferable, and royalty-free license—or something in between the two—adequate for achieving the purposes of the government?

At the present time there is no inclusive single national policy in this area. There is a patent policy established by executive order to cover the inventions of government employees, and there are statutory policies extended to NASA by the AEC covering inventions involving nuclear material and inventions under contracts with NASA, all of which contemplate the taking of title to inventions by the United States, although the Space Act provides discretionary authority both to grant waivers and to issue licenses. In contrast, the statutory patent policy applicable to the National Science Foundation gives a greater discretion to that agency in choosing the extent of rights to be acquired in inventions based upon the requirements of the Foundation and the relative equities involved. With respect to the Department of Defense, there has been established no statutory patent policy, but the policy adopted by defense regulations results in the United States obtaining only license rights to inventions made under defense contracts.

The experience of NASA should be helpful in arriving at a future choice of patent policy for new technologies selected for management. By virtue of the statutory requirements to take title NASA has experienced some unwillingness, and a few instances of outright refusal, to take NASA contracts or sub-contracts. At the very least, this situation can lead to delay in urgent problems. Moreover, the existence of differing patent policies between the military agencies and NASA has resulted in inequities and confusion for industry and government people alike. This has arisen from the fact that in many instances the same general technology is involved under both NASA and military programs; and yet different patent policies must be applied, depending upon whether a military or a civilian space program is involved. Such a situation could give rise to a greater willingness on the part of industry to participate in military rather than civilian programs.

PUBLIC OPINION

In the areas of public opinion and propaganda the problems generated by the space program may be unique. However, the difficulties these problems pose and the possibility of their future relevance suggest that it is worthwhile to discuss these matters here.

If there is doubt whether our country really has faced up to the facts of life in the technological struggle, there is no doubt that when, as in the space program, the United States is made to look second-best, there is an outpouring of public opinion in this country demanding that we regain first place. Moreover, Soviet exploitation of their technological successes tends to result in a downgrading of American prestige in the minds of people throughout the world. This result is no mere accident. There is every evidence that such a result is a principal purpose of the Soviet Union's technological effort, particularly in space activities.

Without a much greater allocation of funds this country could not match every Russian achievement in a contest such as is occurring in space technology. NASA has recognized that it must have some spectaculars along with a broadly based, technically sound program. Without the latter the program would flounder. This underlies the reluctance of top administration officials to state that we are in a space race with the Russians. In attempting to avoid the race psychology, however, there has been the accompanying problem of avoiding the appearance of apathy or mismanagement. The solution has been to acknowledge the intense competition on a broad front with the Soviet Union and to accept the challenge of that competition as being most exciting and visible in the area of space research and development, in which the Soviet Union has demonstrated obvious technological superiority in particular phases of the competition.

There is also the problem of how to translate solid technological achievements into prestige for American efforts when our competitor is playing to the gallery with great effect. Space programs are highly technical. It is hard to convince the general public that discovering the Van Allen Belt may have greater long-term significance than some of the Soviet Union's achievements that have been made possible almost entirely by formidable launch capability. This is a problem intrinsic to programs of science or technology. Their technical nature makes full understanding and appreciation by the public difficult to achieve. It also makes this a field where the public can be misled by the unscrupulous protagonist.

Moreover, there may be a tendency toward conservatism among the members of the scientific and technical community that resists much "tooting of the horn." Obviously, in the long run, achievement and not propaganda will be the key to prestige, and honesty and candor will receive their reward. But in the interim the importance of being able to articulate for the public the technical and scientific achievements of the space program is becoming increasingly apparent to NASA.

Attention must also be given to the manner in which the program that is adopted is exploited in order to project the best image of what we are doing. There must be an effective working relationship between the interested agencies of the government in this area if we are to get the most out of our accomplishments in terms of world opinion. Thus the first United States space exhibit was held in Montreal rather than in the United States, an indication of working together by NASA, the United States Information Agency, and the Department of State.

In this whole picture there are two conflicting pressures. First, is the desire of the press to assure the complete release of all information connected with management of this new and almost completely unclassified activity. This desire has been supported by a specific statutory provision requiring that "information obtained or developed by NASA in the performance of its functions . . . be made available for public inspection" subject to certain narrow restrictions, one of which relates to classified information. There is strong and active Congressional interest in seeing that this policy is strictly adhered to.

The second pressure stems from our need to enhance the prestige of the United States abroad. This pressure tends to favor minimizing failures and maximizing successes. Running through these conflicting pressures is the problem of what information should be classified to protect the interests of the United States. The very nature of space technology has presented NASA with these problems in a very acute form. For example, what should be done with firing dates? If they are given out well ahead of time and then the program slips, there is a loss of prestige for the United States. And yet, it has turned out to be almost impossible to conceal these dates.

PERSONNEL

If less space is devoted to this than to any other major topic of this chapter it is not because of its lack of importance. In fact, it may be the most serious of all the problems facing a government agency whether it is managing major new technologies or otherwise. It is intended here, however, merely to make several brief observations arising from NASA's particular experience.

First, the Space Act authorized a large number of "excepted" positions (excepted from the application of civil service regulations) for which higher salary levels were allowed that tended to be more competitive with the salary levels of industry. Second, as a glamor agency NASA has been able to attract people that might otherwise have been impossible to acquire. If the glamor should wear off, NASA would be in somewhat the same position as other federal agencies except for its favorable salary structure, which is of course quite significant.

The lesson to be learned here is that in the course of framing the basic legislation which establishes the organization, the salary structure must be given the attention it deserves along with any other provisions in the law, namely, top consideration.

It should also be noted that in a space program, where most of the research and development is done under contract, a primary need is for people equipped to manage. This necessarily means people in the higher pay brackets, for it calls for the types of skills for which industry pays very high salaries. It is not surprising therefore that almost all of NASA's staffing has been with people acquired from other government agencies or from NACA rather than from industry.

INTERNATIONAL IMPLICATIONS

One of the principal problems that has faced NASA is that of determining when the technology is ready for international exploitation. Experience to date has indicated that during the early stages of a new technology either domestic or foreign interests will urge the United States to enter into broad government-to-government agreements of cooperation. NASA has been so approached. The Atomic Energy Commission did in fact enter into some broad-scale agreements at a relatively early stage.

The danger inherent in entering too early into broad agreements with foreign countries lies in the disillusionment that may ensue if the promise of economic or other benefits does not materialize relatively soon after. In the case of the Atomic Energy Commission five years have elapsed since international agreements on cooperative programs were reached, but tangible benefits in the form of effective programs are reportedly disappointing. This has the result of making the foreign countries more skeptical of entering into agreements in the future.

An alternate approach to early agreement on a broad scale at the governmental level is to first conduct technical discussions among the respective technical people of both countries. The agreed content of specific technical programs may thn determine the content of viable formal agreements. NASA has been following this alternate approach.

Dangers notwithstanding, some benefits may accrue to the United States from international programs. Of course, certain foreign policy objectives are served. Beyond this, however, there may be tangible benefits in a more economical and effective utilization of scientific and technical personnel, facilities, and know-how when the resources of other countries are tapped. This may result in both a saving of valuable and scarce American resources and the avoidance of possible duplication. Moreover, development of the science or technology involved may benefit directly from the participation of foreign countries, to the advantage of all concerned.

The international implications of placing a military label on this program were alluded to earlier. In the management of the space program it has been a critical consideration. Since spacecraft lend themselves to such military purposes as reconnaissance, some foreign countries understandably have needed reassurance that the United States program to explore space is not military. If the United States had not been able to provide such assurance, it might have been impossible to negotiate foreign rights in a number of countries for construction and operation of tracking stations. This conclusion is supported by the extreme sensitivity shown by those countries to any possible military implications to the space projects involved. Even the question of using military men to be Project *Mercury* astronauts was given careful study.

An additional example of the unacceptability of a military label is the proposal that a space program be established under NATO auspices. European scientists had no enthusiasm for this proposal because NATO to them was already a military organization and to attach such a stigma to something ostensibly devoted to the exploration of space for peaceful purposes was unacceptable. Such a proposal might have proved more acceptable if the image of NATO had been successfully transformed to other than that of a purely military arrangement.

A final aspect of the international side relates to the manner in which the managing agency should work with the Department of State and foreign countries. To what extent should foreign policy objectives play a part in the formulation of the program? Should the managing agency just keep the State Department informed or should the latter play a more active role?

There is also the problem of the extent to which formal government-to-government agreements, as distinguished from agreements reached below the top governmental level, are required in entering into cooperative agreements in foreign countries. The arrangements that NASA has entered into have ranged all the way from contracts between NASA and a foreign university, in which the State Department did not participate, to formal agreements at the governmental level in which negotiations were conducted by the Department.

Conclusion

In conclusion it seems desirable to take a somewhat broader look at the whole matter of new technology, viewing it from the standpoint of its implications for our current way of life. Behind these concluding comments lies a question that should be the concern of American policy-makers. Are we prepared as a nation for the challenge that a major new technology poses for us, or, at least, are we taking real steps to be prepared?

What, then, are the broader implications of new technology? What is the challenge for which we must prepare ourselves? One major challenge was discussed in the introduction to this paper. It is the fact that we are engaged in an across-the-board competition with the Soviet Union in which a key element is the technological race. Our chief competitor is skilled and intent on winning. In his mind the stakes are the ultimate domination of the world by the Communist system of government.

On the sociological and economic plane also serious challenges may be thrown down by new technology. One needs merely to let the imagination wander to realize the possible implications—for example, that basic precepts held by many as to the role of government in our capitalistic economy could be significantly affected as a result of technological developments.

The principal point intended to be made here is that in the course of the next twenty years some very fundamental elements of our way of life may be, or may appear to be, seriously challenged by new technology. The Soviet Union is not bound by the same principles as we are—although it may be a prisoner of Communist ideology. Are we preparing ourselves to meet this challenge in a manner that will advance the interests of the United States and carry out our national goals? Will we be flexible and imaginative enough to see the opportunities of new technology for the benefit of mankind, or will we tend to run from this challenge because it appears to threaten the things we hold sacred? Will we use our energy to shape the course of the future in new technology, or will we find ourselves helpless as a nation to rise to this challenge?

We may not be able to count too heavily on the inflexibility of the Soviet Union as a prisoner of its own ideology. Soviet leadership has shown itself quite adept at rolling with the punches and adjusting Communist ideology to meet the times. Are we taking the steps today that will be necessary to establish the fact that a democracy which holds sacred the ideals set forth in the United States Constitution is at least as well equipped to deal with the challenge of the future in the field of new technology?

5

International cooperation in space science

♦ Hugh Odishaw

The opportunities and problems of international cooperation in space are many and complex. Among the opportunities are those very general ones associated with a vast hitherto inaccessible region—a virgin territory to which man now first turns. Some insights into space in this sense may be had by looking briefly at Antarctica and at atomic energy. Additional insights may be had by examining the record of cooperation among scientists within their own nongovernmental structures: the history of the International Council of Scientific Unions (ICSU), the essential qualities of the International Geophysical Year (IGY), and the evolution of the rocket and satellite program of the IGY and of the ICSU Special Committee on Space Research (COSPAR). While the non-

105

♦ HUGH ODISHAW has been Executive Director of the National Academy of Sciences' US National Committee for the International Geophysical Year since 1953. He is also Director of the US IGY World Data Center, Executive Director of the Space Science Board, and Executive Director of the Geophysics Research Board. His books include the *Handbook of Physics* (with Dr. E. U. Condon) and *Science in Space* (with Dr. L. V. Berkner), and he is a recipient of the Navy's Distinguished Public Service Award.

governmental role in space has been quite prominent, it is worthwhile to consider the progress that has also taken place in bilateral and regional cooperation among governments. In the light of these developments one can then turn to prospects and problems before us today, involving such issues as science and politics, cooperation and control, and the future of nongovernmental collaborative activity.

The Qualities of Space

ANALOGY OF ANTARCTICA

Antarctica is a desolate, snow-and-ice-covered continent of almost six million square miles. It is now by treaty devoted to peaceful purposes and is the only portion of Earth so dedicated. Yet claims had been made to most of it; conflicts among Argentina, Chile, and the United Kingdom had erupted from time to time; and the interests of both New Zealand and Australia in the waters and terrain below them have been keen. That a treaty was possible is attributable in large measure to the fact that the continent is uncongenial. Its only export is scientific information largely about weather and climate, the earth's magnetic field, the ionosphere, the aurora australis, cosmic rays, glaciology, seismology, and gravity. Its vastness and the rigors of its environment make difficult even the conduct of scientific observations on any sizeable scale.

The first great attack upon the scientific problems of Antarctica occurred as part of the IGY program, remarkably successful both in its results and in its spirit of cooperation, which established the foundation for the treaty; and the work goes on as an extension of the IGY under similar international scientific auspices.

Space also is vast and desolate, and its environmental rigors make Antarctica an Eden by contrast. Its population of electromagnetic fields, particles, and radiations has no palpable appeal to nationalism or power

106

politics. Its contents are largely beyond the reach of space vehicles, and even the nearer bodies, lying within our solar system, have less appeal than Antarctica as spheres of human activity. Thus the essential bareness and recalcitrance of space, far exceeding the analogous characteristics of Antarctica, provide no substantive arena for power politics. Yet space affords a vast challenge in the pursuit of knowledge about the universe analogous to the opportunities provided by Antarctica for rounding out our knowledge of the Earth's physical nature.

Space—near and far—appears to be favorably constituted for cooperation. Far from the Earth, no real prospect for friction among nations exists, and progress in science would benefit from cooperative and even collaborative approaches. Near the Earth, the new tools raise problems relating to man's control of his environment: cooperation among scientists to assess these would lay a technical foundation for intergovernmental regulation which sooner or later will be forced upon the nations of the world. Meanwhile the planetary and interplanetary character of the geophysical sciences affords a basis for the continuity of cooperation among scientists.

ANALOGY OF ATOMIC ENERGY

For good or ill, our century may be remembered best for two accomplishments: harnessing of nuclear energy and rendering accessible the far reaches of space. There are differences between the two events, but looked at from afar both have much in common. Each development embraces both science and technology, for each is linked to activities and interests that range from the pursuit of knowledge for its own sake to the development of practical devices and services. Each has both beneficent and forbidding aspects. Both are linked to the issues of war and peace, and both have engendered opportunities and difficulties in the relations of men.

Nuclear energy was born of war, and its initial product was a weapon of unprecedented destruction, quickly caught up in power struggles. The space age, on the other hand, was ushered in during peace—however sharp the political dissensions of the period. Further, it was ushered in under the aegis of an unprecedented program of international cooperation, the International Geophysical Year. This is a distinction in both origin and development, and it could be a significant distinction in the way the space age fares. Like nuclear energy, space has been linked to warfare, but the linkage stems simply from two independent applications: rocket engines used in space systems are similar to the engines used in guided missiles, and satellite payloads can fulfill a military as well as a peaceful purpose. The engines of bombers and commercial aircraft, of battleships and ocean liners, of military trucks and passenger cars are also similar in the same way. Yet these similarities are not sources of misunderstanding, and there is no substantive basis for a different attitude toward rocket engines per se.

Space accomplishments nevertheless reflect the capability of a nation's rocket engines, which when translated into missile capability gives rise to the problem of competitive status in weapons. Such comparisons are part of what has become a far more general question: the actual and symbolic scientific and technological status of a nation as a factor in power politics. Nations therefore seek to assert their leadership in science and technology. Such assertions are highly competitive, but competition here is preferred to competition in battle. Moreover, competition does not always preclude cooperation, particularly where there exist areas of activity that transcend both national and even planetary bounds.

There is, then, some basis for hope that space can be characterized by peaceful exploitation and by cooperation even if an appreciable element of competition exists. If so, the brief and limited tradition of cooperation among scientists in space research may prove to have been helpful although the scientific community itself will have to contend with problems as well as opportunities. Cooperation among scientists, however, depends not only on their own actions but also on those of their governments and of intergovernmental bodies.

Cooperation in Geophysics

HISTORICAL

Some insight into the problems and prospects for cooperation in space may be gained from an examination of what has happened and is happening in the scientific community. What is relevant is not only the short history of cooperative effort in space science but also the longer history of cooperation in geophysics.

In the nineteenth century three fields of geophysics called for international cooperation: the determination of precise latitudes in view of the Earth's shape, the delineation of the Earth's magnetic field, and the timely accumulation of weather information. Joint efforts in both meteorology and geomagnetism were somewhat formalized through institutions in Germany. World War I, however, put an end to this cooperation. After the war, while resentment of Germany was still high, the academies of several countries organized an International Research Council for cooperative purposes but excluded the Central Powers. This exclusion led to the decline of the Council, for it was recognized that there could be no truly worldwide cooperation while several groups of scientists were barred, and, more important, that political considerations had to be eliminated if an international scientific organization were to be effective. Accordingly, the International Council of Scientific Unions was established without political restrictions as to membership. The Council had dual membership: scientific institutions throughout the world—primarily academies or research councils—and the subject-matter international unions such as those devoted to geophysics, astronomy, physics, and chemistry. These unions themselves were composed of scientific representatives from the various research institutions throughout the world.

The Unions of the ICSU number fourteen and deal with astronomy, geophysics, geology, chemistry, radio waves, physics, biology, physiology, biochemistry, crystallography, mechanics, mathematics, geography, and the history and philosophy of science. They provide a forum for meetings and discussions, publish leading international journals, and conduct symposia between their triennial assemblies. The Unions have provided a most valuable service to science and society in their definition and standardization of nomenclature, fundamental physical constants, and units and physical standards of measurement. Moreover, they establish special committees as necessary for studying specific current topics.

The problem of initiating and conducting major cooperative enterprises on the international scale raises the question of the effectiveness of committee action. The IGY served as a useful example of how such a committee, if thoughtfully conceived, based upon genuine needs, and composed of experts who themselves are participants, can combine the enthusiasm of individuals with the necessary structure for coordinated activity. The origins of the IGY themselves reveal how effective individual human beings can be in this way.

THE IGY

The initiation of this vast enterprise had simple beginnings. In April 1950 a small group of geophysicists, including the American, Dr. L. V. Berkner, and an Englishman, Sydney Chapman, were at the home of Professor J. A. Van Allen in Silver Spring, Maryland. The discussion turned to the status of geophysics, and Berkner proposed the conduct of a worldwide observational and experimental program as a means for stimulating research and, with the aid of modern tools, markedly accelerating the rate of progress in geophysics. During the next two years Berkner and Chapman presented this proposal at meetings of several international bodies concerned with the field of geophysics. In the course of these discussions, which met with enthusiastic response among the world's scientists, the definition of the IGY was developed.

To afford a forum for detailed planning and to coordinate the efforts of sixty-six countries, the International Council of Scientific Unions established in 1952 the Special Committee for the IGY—known in short as the CSAGI, after the French initials of its name (Comité Spécial de l'Année Géophysique Internationale). The first meeting of the CSAGI in 1952 produced an outline of the general criteria for the program. There followed a succession of IGY assemblies in the course of which hundreds of specialists from all parts of the world gathered together during the next four years to spell out the detailed program and to work in national contributions. It was through these general assemblies of a nongovernmental body that the planning and coordination of the IGY was achieved, supported by a small and temporary secretariat in Belgium.

This development has two aspects of significance. First, it reveals that the efforts of a few creative individuals, assisted by the basically informal sponsorship of existing nongovernmental scientific organizations, were able to effect a major human as well as scientific enterprise. Moreover, it suggests the importance of the role of individuals themselves in scientific affairs. Second, the IGY demonstrates the ability of nongovernmental scientific organizations to plan and conduct a large-scale collaborative effort—an effort which was dependent upon its spirit of harmony and friendliness. The latter, in turn, while stemming in part from the objective qualities of the subject matter, was possible only because political considerations were essentially absent, and absent because irrelevant. Governmental organizations, however anxious to achieve such ends as sought by CSAGI, tend always to be handicapped by the total political context of their members' inter-relationships, making it difficult if not impossible for them to exclude political factors whether relevant or not.

The objectives of the IGY included space science because studies of the upper atmosphere and solar-terrestrial relationships bulked large. But it was 1954, more than four years after the initial proposal, when the rocket and satellite program was incorporated into the effort. For not until then did it appear that the state of rocket technology provided real hope for satellite ventures. With expectations aroused by postwar rocket engine developments, and anxious to secure information on time and space variations of particles, radiations, and fields near the Earth, the scientists participating in the IGY General Assembly at Rome in October 1954 recommended that "thought be given to the launching of small satellite vehicles."

What were the significant characteristics of the cooperation in the rocket and satellite program of the IGY? In general, it was characterized by the over-all remarkably harmonious nature of the IGY. As in other areas, a working group was established, providing a forum for discussions and communication among its members, all of whom were creative working scientists. Specifically, one can cite several examples of cooperation in space during the IGY. Program reports were interchanged and discussed, although the Soviet satellite presentations were very general. Rocket flight summaries were interchanged. Tracking information for satellites was issued, although the Soviets preferred the medium of the press to the IGY international communication network and although the Soviets provided station predictions rather than the more general orbital elements desired by Western scientists. Research results were interchanged through the IGY world data centers. Soviet results flowed more slowly; they also were not as rich as those of the United States, but this deficiency stemmed from the nature of their program which, in my opinion, stressed historical "firsts" and exploration, made possible by their larger-thrust engines, in contrast to the much broader scientific program of the United States conducted aboard smaller but more plentiful vehicles.

Yet there were problems, some springing from the difficulties of the new field itself, some from the very fact that only a few nations possessed the new tools, and some probably political in origin. Thus national interests did enter into the space picture, even during the IGY. In the space program certain aspects of national interests focused on the gross achievements of space systems—i.e., the capability of rocket engines. Science itself became a peripheral consideration despite the language used in statements originating in both the Soviet Union and the United States. Perhaps government investments were too great to avoid capitalizing upon space feats, and the fact that only the United States and the Soviet Union had satellite capabilities heightened the competitive element.

At the time of the IGY, sounding rockets—research rockets used to obtain data on the upper atmosphere—were available to a few nations: the United Kingdom, which collaborated with Australia; France, which did not undertake an IGY rocket program; Japan and the Soviet Union; and the United States, which also worked closely with Canada at the joint Fort Churchill rocket launching site. But only two nations possessed satellite launching capabilities; and lack of space tools obviously makes it difficult to bring into play the interests and enthusiasms of scientists who do not have the tools. These restrictions are slowly but surely fading because the tools are now becoming more generally available, space data are attracting more analysts, and ground-based studies are becoming significant adjuncts to space experiments.

CHARACTERISTICS OF IGY

The essential positive features of the IGY, viewed as a major exercise in collaboration in science, were several: its harmonious, enthusiastic spirit, its simplicity of organization, its largely apolitical nature, and at the same time its appeal to national interests.

Its spirit had its source in two attributes of science. The objectivity possible in science, rooted in the objective, universal quality of nature, is conducive to objectivity and harmony in relations among scientists. Specialized competence in research breeds enthusiasm if problems of a genuine scientific nature are present. The formulators and planners of the IGY were the world's leading workers in their several fields; the problems of geophysics were real, and their solution did call for international collaboration; moreover, the planners and the doers were the same men.

The organization was never cumbersome since the international secretariat never exceeded a few people who were themselves working scientists. Thus there was a simplicity and directness in administrative problems and relationships. This was achieved in large measure by delegating many of the coordinating tasks to experts in each field of geophysics, while many problems of communication and administration relating to national problems were assigned to the national committees themselves.

Because the organization was not a governmental body but a body of private citizens, it was possible for it to be almost totally apolitical. The few political problems that did intrude were successfully coped with, and their intrusion was not allowed to affect the over-all program as it was conceived.

Non-governmental and apolitical, the IGY nonetheless succeeded in eliciting the support of nations—perhaps largely because the scientists of each nation presented the program to their own governments. Their governments responded and generously supported their proposals, in part because the program was thoughtfully conceived of by the world's leading scientists, in part because the submission was made by a nation's own experts, in part because there were national values to be gained from increased research and national prestige itself was involved. Furthermore, the international cooperative aspects of the venture themselves had appeal to governments as well as individuals.

The combination of these elements, plus the stimulating catalytic effect of the whole enterprise as it evolved, not only accounted for its success as it was conceived but also led to a marked augmentation of geophysical research throughout the world and to marked interest in the venture by peoples everywhere.

Cooperation Through COSPAR

One indication of the value of the cooperation of the IGY was the emergence of successor enterprises in the nongovernmental realm. Of these, the establishment of the Committee on Space Resarch (COSPAR) is most pertinent. Under ICSU auspices, space cooperation has progressed along the lines of the IGY. New problems have arisen, but they have not inhibited an extension of the IGY pattern.

COSPAR itself did encounter political problems in the formulation of its own charter, initially established by the International Council of Scientific Unions at its general assembly in September 1957. The membership of COSPAR was established as consisting of national members representing academies of sciences and of union members representing the scientific unions of the Council. The national members were to come from countries active in rocket and satellite work; these were Australia, Canada, France, Japan, the Soviet Union, the United Kingdom and the United States. In addition, three other members were to be drawn from countries active in tracking or other aspects of space research; these were to be rotated among the qualified countries. Lastly, nine of the ICSU unions were to be represented.

COSPAR's initial meeting took place at London in November 1958, and a charter much along the Council lines was adopted. At the second meeting of COSPAR, March 1959, the Soviet member proposed the addition of the Ukraine and Byelorussia to the category of countries launching satellites or rockets and six other Eastern European countries to the category of rotating membership. This proposal was not adopted, and the future of Soviet participation became uncertain by the close of the meeting. There was general acknowledgment, however, that for a truly international effort within COSPAR the way should be open to more countries interested in participating in the work of COSPAR, and the matter was referred to the International Council of Scientific Unions. In June 1959 the charter was revised to permit any national group adhering to the Council to participate in the work of COSPAR. This modification was accepted by the Soviet Union as the basis of discussion.

The Charter as subsequently approved provided for this modification, but it also contained what has been called a veto power for the Soviet Union and the United States. In point of fact, however, the veto power is neither complete in any formal sense, nor meaningful in any ultimate practical or operational sense. The latter will be discussed below; the former, as a topic of organizational interest, may be explained now as follows. The veto power exists in the Executive Council (composed of the COSPAR Bureau and the representatives of the ICSU scientific unions) because decisions of this body must be confirmed by a 2/3 vote within the Bureau (composed of an elected president and two vice presidents, and four additional members; the United States and the Soviet academies may each control the nomination of one vice president and two members). But no veto power exists in COSPAR, which includes both the Executive Council and all representatives of adhering scientific institutions. However the definition of functions of COSPAR and its parts (the whole, the Executive Council, the Bureau) and the provisions of by-laws establish control, by the United States and the Soviet Union, largely on two matters: financial assessments on adherents in support of COSPAR, and relations with the UN on "regulations affecting space research" (but not on other relations).

COSPAR has been successful in continuing and even extending the IGY pattern of cooperation in space. Annual programs are gathered and distributed, with their content improved. A space bulletin, limited thus far in content, has been initiated and holds promise of serving as an authoritative timely medium for international communications. Orbital elements are now accepted as the means for providing tracking information, and interchange of results continues through the world data centers. An annual series of Rocket Weeks has been established, and special topics have been studied, including space needs for radio frequency allocations, the properties of the upper atmosphere, and certain annual geophysical events associated with solar activity and the upper atmosphere. Annual general assemblies have been initiated which include symposia providing scientists with an exceptional opportunity for discussion as well as presentation of research papers.

The cooperation within COSPAR is benefiting from several developments. First, interests in space science are increasing, and some fourteen nations have established scientific committees[1] concerned with space which are working with COSPAR. Second, the tools of the space age are becoming more broadly available. Although the United States and the Soviet Union are the only two nations at this time with satellite capabilities, the United States is making available vehicles to experimenters of other nations. Sounding rockets, simpler and less costly, are being used by more nations. Moreover, active experimental work in many other nations affords a broader base for a truly international approach to space science. Third, the results of space science so far have vitally affected all scientists interested in the upper atmosphere and space. The availability of data has permitted analysis and theoretical studies by scientists in many parts of the world, and the utility of space tools in gathering new data, now proven, has stimulated scientists in even classical fields of inquiry.

The results obtained from satellites and deep space probes have led to a re-evaluation of theoretical and experimental work carried out on the Earth itself in such fields as optical and radio astronomy, ionospheric physics, geomagnetism, cosmic rays, and auroral physics. Many nations have scientific groups active in these fields. The significance of these activities has been heightened by space results because ground-based work can contribute directly to the solution of these same problems. For example, the recently intensified interest in the reaching of the Moon and the nearer planets with manned or non-manned space systems has forced attention on planetary astronomy. Yet during this century the interest of

---

[1] Space committees have been established in or by Argentina, Australia, Belgium, Brazil, Canada, Czechoslovakia, France, Italy, Japan, Netherlands, Sweden, Switzerland, United Kingdom, United States, and the Soviet Union. Some of these committees are governmental, some are not. Adherents to COSPAR, usually the academies or research councils in various countries, include representatives from such bodies in Argentina, Australia, Belgium, Canada, Czechoslovakia, France, the German Federal Republic, Italy, Japan, Netherlands, Norway, Poland, South Africa, Switzerland, the United Kingdom, the United States, and the Soviet Union.

astronomers has been devoted to galactic and extra-galactic astronomy. It is apparent that much can be learned about the Moon and the planets by intensified ground-based programs which would provide scientific knowledge perhaps crucial to space system ventures and, in any case, intrinsically valuable.

Bilateral and Regional Cooperation

To some extent the lack of widespread rocket and satellite tools has been compensated for by bilateral arrangements entered into by the United States Government with foreign governments. (There is no indication that the Soviet Union has engaged in such activities.) The United States has entered into arrangements with several nations. This growing program of the National Aeronautics and Space Administration is based on its enabling legislation which states that

> the aeronautical and space activities of the United States shall be conducted so as to contribute materially to . . . cooperation by the United States with other nations and groups of nations in work done pursuant to this act and the peaceful application of the results thereof.

Although the NASA program does not involve the interchange of funds, it is rooted in truly joint efforts which vary in their nature, depending upon the characteristics of the scientific problems. Careful discussion of proposed experimental work at the scientific working levels always precedes formal negotiation. NASA has committed to joint programs space vehicles for satellite experiments, room in space vehicles for other experiments, sounding rockets, and rocket research instrumentation. Under such arrangements both England and Canada prepared satellite payloads for launching by NASA in 1962, while France worked out specific proposals for a French-United States satellite. The Italian space community and NASA developed a research program in which Italy procured United States sounding rockets and provided most of the ground-based equipment while NASA provided the payload and basic launcher. Other similar programs have been initiated with Norway, Sweden, and Argentina, and additional programs are understood to be in process of negotiation. Technical advice is provided by NASA when needed in these joint efforts, and close relations are maintained between appropriate groups of scientists.

Other promising and useful aspects of the NASA bilateral program include the ground-based observation programs relating to experimental communications and weather-forecasting satellites. This joint participation and that involved in the global satellite tracking network have stimulated scientific interests abroad. Several types of cooperative training and fellowship arrangements entered into by NASA with foreign scientists have also contributed to international scientific interests and cooperation.

One of the most significant aspects of this bilateral government program has been its operation within the non-governmental COSPAR framework. It was at a COSPAR conference that the National Academy of Sciences presented a NASA offer to COSPAR members to launch specific experiments or complete satellites. As in the IGY, COSPAR does not undertake to negotiate bilateral arrangements. It does, however, encourage such arrangements and provides a congenial forum for the discussion of experimental ideas. The presentation of such offers within COSPAR in itself creates an international environment for the experiment and for the scientists. This, in turn, is an important factor in keeping the experimental work that may ensue within an international framework not only for discussion purposes but also for the interchange of data and results.

The NASA procedure compares favorably to that in the analogous area of atomic energy. Although the International Atomic Energy Agency was not established for cooperative scientific efforts, it was authorized largely to assist underdeveloped countries in applications of nuclear energy, to conduct the sale of uranium fuel under safeguards as to their possible military diversion, and to develop safety standards. But even before, as well as after, the agency was established both the Soviet Union and the United States entered into arrangements for the sale of uranium fuel at costs below those which the Agency could meet, arrangements which are wholly bilateral in nature and seriously hamper the ability of IAEA to promote international agreements in atomic energy. NASA, on the other hand, has endeavored to establish its bilateral arrangements within the international framework or spirit of COSPAR.

Bilateral arrangements are not the only approach to the problems associated with increasing the availability of tools and facilities for space research. Regional programs also offer much promise in broadening the base of active rocket and satellite work. The first major attempt to this end is already under way in Western Europe where discussions have proceeded along two lines: first, the establishment of a central facility for theoretical and experimental activity, and second, the establishment of a common Western European launching capability. The intent is the development of activities similar to CERN, a western European enterprise in nuclear physics with facilities, including a large high-energy accelerator, at Geneva. CERN has proven to be a remarkably successful venture as a productive research facility. Although Western European governments formally entered into arrangements for this facility and support CERN, the administration is simple and flexible and is in the hands of competent creative scientists. The conduct of the research and of scientific relationships falls within the traditions and customs of the scientific community.

Prospects and Problems

While cooperation in space science during its first few years of existence has been productive, the opportunities in the years ahead promise to be so vast that it is of interest to consider the prospects of continuing and increased cooperation. Crucial to the extension of cooperation within the scientific community are two factors: permissive attitudes on the part of governments, and the ability of the scientific community to plan and conduct major efforts.

SCIENCE AND POLITICS

Rightly or wrongly, it appeared to scientists involved in COSPAR's deliberations that the Soviet reservations on the COSPAR charter were politically motivated and, moreover, that they did not represent the convictions of Soviet scientists. We have noted that after Soviet failure to obtain membership for two republics of the Soviet Union and six Eastern European nations the veto element was injected. That action suggested to some the prospect of future Soviet moves in other nongovernmental agencies as well as the introduction of new political elements in the deliberations of the scientific community. Thus far, such Soviet moves have not materialized.

Although the scientific community considers any veto provision in the COSPAR charter unfortunate, it is not construed as crucial, largely for two reasons. In the first place the provision that permits the academies of the United States or the Soviet Union to exercise control applies in the main, as to specifics, only to financial assessments levied on participating national groups in the support of COSPAR and in relations vis-a-vis the United Nations. Moreover the provision in its generality applies only to the Executive Council. It does not apply to COSPAR as a whole, and for the reason some argue that in any ultimate, formal sense there exists no real veto provision at all: COSPAR as a whole, which includes 28 union and "national" adherents, operates on majority vote—usually a simple majority, because only financial assessments and changes in by-laws go beyond a simple majority to a 2/3 vote. Thus, in view of the membership of COSPAR, neither the U.S. nor the Soviet academy exercises control.

In the second place, and operationally far more important, cooperation among scientists is not a matter of vetoing or even voting, but depends on general agreement and enthusiasm for an activity—almost in the spirit of the Quaker "sense of a meeting." Moreover, some scientists of Western Europe felt that the special circumstances—namely that only two nations had launching capabilities which involved enormous costs—warranted for the time being some added role for the United States and the Soviet Union. For such reasons assent to an apparently larger voice for the academies of the United States and the Soviet Union took place. It did permit Soviet scientists to participate in COSPAR—and this was the vital factor.

In any case, the veto episode is significant because it revealed that political or nationalistic considerations can be injected into the scientific community. It pointed to the fundamental requirement that governments must allow their scientists to work together in international bodies on appropriate scientific problems in the traditional spirit of scientific cooperation uncolored by political considerations. Thus far, fortunately, the activities and deliberations of COSPAR have proceeded along the usual lines of scientific intercourse, and it appears that COSPAR can now turn more aggressively to furthering cooperation in space research.

FUTURE OF COSPAR

First, no real problems lie before COSPAR in continuing the cooperation which characterized the IGY—cooperation that takes the form particularly of sharing the discussion of and plans for experiments, the conducting of symposia, and the interchange of results. COSPAR can continue to encourage discussions of bilateral and regional scientific interests. It can serve to represent the needs of research on issues of regulatory nature, as it has done on the problem of allocation of frequencies for space vehicles. It can continue as a mechanism for personal relationships and for the submission of individual ideas and proposals. It is, in fact, this element which is least understood outside the scientific community and which is the crux of international cooperation in science as typified so strongly by the IGY.

Second, COSPAR can consider conducting major efforts approaching and perhaps even exceeding the magnitude of those of the IGY. It has already established the Rocket Week intervals, which may turn out to be a very important and expanding feature of near-space research. The increasing availability of space tools is also significant for potentially greater cooperative efforts.

The fact that only two nations possess satellite launching capability has appeared restrictive, but this situation, as we have noted, is changing. Sounding rockets are becoming more generally available not only through bilateral arrangements but also through the initiative of individual countries. Israel has recently developed a sounding rocket on its own initiative; and during the IGY Japan successfully designed and produced small research rockets. Moreover, the range of sounding rockets will also develop in time; they will have considerable use not only in sounding the upper atmosphere but also in going out much further. In time these technological developments will lead to satellite capabilities in other nations besides the United States and the Soviet Union.

Moreover, the data from space systems can be used by scientists everywhere. Already an appreciable amount of research using such data has been done by scientists in non-launching nations. These opportunities will steadily increase, and they may be valuably augmented from time to time by satellites whose instruments transmit continuously. To a scientist anywhere this will be as though the instruments package were his own: he can receive and record the telemetered data, reduce it, interpret it, and publish his results.

Third, with this steadily expanding reservoir of interested scientific groups, skills, and tools, COSPAR can begin to look into problems of science that would profit from international cooperation and collaboration. Now that the IGY data have been sufficiently digested, the time is ripe for new efforts. As a matter of fact, two such efforts have already been initiated: the World Magnetic Survey, and the International Year of the Quiet Sun. Both will take place during the coming sunspot minimum, 1964-65, and both will be concerned with phenomena of the upper atmosphere, near space, and solar-terrestrial relationships. Both have proposed experiments that call for space tools.

While these enterprises are being considered by another committee of the ICSU—the International Geophysics Committee (CIG)—COSPAR is playing a leading role in those aspects that embrace space science. Once the programs of this period are under way, COSPAR can turn to the future. With its normal cooperative work maintained year by year, it is quite likely that COSPAR can from time to time, as the needs of reseach dictate, undertake special intensified efforts. In all probability these efforts will be concerned with the problems of the solar system, and here solar activity suggests a reasonable timetable. The sunspot cycle runs for some eleven years in duration. Solar activity minima and maxima are unusually interesting periods for investigations of electromagnetic fields, particles, and radiations. There would be logic to the conduct of intensified concerted efforts at these times.

COOPERATION AND CONTROL

The challenges implied in the exploration of space are not restricted to science. In science itself, in addition to the many problems that are largely restricted to the context of pure research, there are some experiments which impinge directly upon social and political problems. The applications of science, as in communications, weather forecasting, and navigational satellites, touch directly upon social and economic affairs. Already the technology of space systems represents a major industrial activity, and its by-products will have effect in other industries (as Leonard Silk's chapter reports). Moreover, the actual or potential use of space tools for military purposes raises vital questions having to do with world politics.

For instance, it seems clear that only the United Nations, as the highest international body available, could be responsible for such regulatory problems as would be posed by a treaty banning weapon applications in space. But should the United Nations itself also attempt to undertake cooperative programs in space science and technology? It is not certain that it should. First, it is not certain that a single agency can effectively pursue dual objectives—i.e., objectives that are essentially political in character and others that are not basically political. Secondly, the very nature of an organization devoted to political affairs connotes a membership and character that are not most conducive, for example, to cooperation in research—or in poetry or painting. Moreover, the diversion of the interests of such an intergovernmental agency can lead to neglect of its primary, in this case political, function of dealing with the most pressing problems of our age. Thus the greatest contribution that the UN could make to space would be to encourage its member nations in executing a treaty on outer-space control, which would create the world political climate most conducive to international cooperation in all areas of human welfare.

In the years ahead the competitive development of service satellite systems for weather forecasting, communications, and navigation may well represent a serious waste of human resources and may, through confusion, inhibit their most effective, world-wide use. Perhaps intergovernmental organization can be effective in this area; perhaps the scientific community itself may be able to work out appropriate cooperative arrangements. In any event, careful thought during the few years ahead, when initial steps are taken with experimental systems (steps that may further elucidate the problems and the prospects), needs to be given to the nature of these questions and their possible resolution.

While most experiments in space do not in themselves imply problems of regulation and control, there are some possibilities now known, and still others may turn up, which call for careful analysis. Present space experiments largely relate to the upper atmosphere and near space. They are made possible by the fact that space tools now permit man to inject materials into near space which might affect his physical environment. Weather control, far in the offing, represents perhaps the most spectacular of these possibilities, and the one most critical to mankind. Even preliminary experiments, however limited in scope, will probably engender critical reactions by men and nations.

Sooner or later such projects will call for an appropriate international organization. Although it may turn out that they can be coped with in part and during the early stages by the scientific community, an intergovernmental approach may be not only preferable but mandatory. The United Nations itself, confronted with highly controversial and pressing political problems, does not appear to this author to be an ideal forum; no UN specialized agency appears to have the appropriate authority; and it may be desirable to consider the nature, structure, and authority of a new specialized agency concerned solely with regulation of all activities that seriously affect the gross physical environment of man.

In all of these matters concerned with science or its applications the scientific community can itself be helpful even where it cannot extend its activities into the area of regulation and control. Specifically, appropriate initiative within the scientific community in the technical analysis of such problems, as long in advance as possible and certainly long before their realization, would afford a proper forum for objective discussion. Moreover, it would lead to the existence of a body of scientific and technological data and information, objective and apolitical in nature, to which governmental organizations could turn—as, to some extent, for example, the International Telecommunication Union now turns to the nongovernmental International Scientific Radio Union, a pattern which merits further exploration by both the scientific community and the intergovernmental community. In some ways the burden falls first and most heavily upon the scientific community in anticipating trends and possibilities in science and in its applications which may involve broad human interests, for scientists are likely to be first to know and sense these.

Looking back, we can see that in the period following Columbus' discovery of America the excitement and wonder of his discovery were apprehended broadly and generally, influencing almost every aspect of life. But they were not comprehended well enough to anticipate even fractionally the Americas of today. Looking even further back, we can see that the Copernican astronomy not only achieved a revolution in science but also changed man's concepts of man and of earth. The onset of the space age affords the possibility of a comparable impact.

It may be argued that the new astronomy of Copernicus had its large effect because it was a revolutionary scientific discovery while space is not a revolutionary discovery: it has been there all along and all that is new is the availability of engines that can travel there. To this argument let me give three answers.

First, since space has no meaning unless realizable, its now being tangibly within reach has large significance. To be at the threshold of an age which releases man from the confines of his ancestral planetary home is no small matter.

Second, the power of space tools cannot be denied. Their application in communications and weather forecasting cannot fail to be very appreciable, economically and socially. Their application to controlling or altering man's physical environment is bound to come, whether in one or ten decades.

Third, the very fact that the results of man's coming space ventures are still unknown only furnishes added intellectual provocation. The history of science shows that major undertakings, whether based on new and powerful ideas or on new and powerful tools, are rich not only in new explorations into nature but also in the varied by-products that follow in their wake.

In short, the diversity and scope of space and all that it connotes suggest that fresh and imaginative insights are needed. There are primary political problems to which the United Nations should return anew. There are secondary problems of regulation that existing and new intergovernmental bodies must contend with. There are opportunities for collaboration among scientists along the lines of the IGY, and there are other opportunities for bilateral and regional cooperative endeavors. The danger is that the variety and extent of the impacts of space will be underestimated.

6.

Arms and arms control in outer space

♦ Donald G. Brennan

Few subjects are as complicated and at the same time as speculative as the future development of military systems in outer space.

The difficulties do not stem from uncertainty concerning the likely evolution of basic space technology, such as rocket boosters and guidance systems, the progress of which is already fairly certain even though the rate at which it may proceed is subject to wide variations. Rather, they arise from the fact that the applications of the basic technology to future military systems depend critically on political and military decisions not yet made.

The present paper outlines some of the possible military developments that have been discussed for outer space. (By "outer space" is intended the region above the sensible atmos-

123

♦ DONALD G. BRENNAN, a mathematician associated with the Massachusetts Institute of Technology, is editor of the recent volume *Arms Control, Disarmament, and National Security*. He served as co-director of the American Academy of Arts and Sciences 1960 Summer Study on Arms Control, and has been a consultant to the Advisor to the President on Disarmament, to the President's Science Advisory Committee, and to the Department of State.

phere; ballistic missiles are not considered as weapons of outer space within the terms of this discussion.)[1] It will also discuss some of the arms control arrangements that have been proposed to date, with emphasis primarily on weapon and control possibilities rather than on recommendations. Moreover, the discussion will center almost entirely on possibilities already proposed or under active development at the time of writing.

The Strategic Background

SPACE AS A COLD WAR ARENA

It is worth stressing at the outset that the competition in space technology generally, and in its military applications specifically, is one of the aspects of the cold war and cannot be divorced from that setting. As such, it requires an understanding both of the political setting imposed by the cold war and of the framework of military strategy within which the cold war in its military aspects is being carried on—and may, unless checked, be carried on in space.

Although Soviet officials have not said much explicitly about their present and projected military capabilities in space, they possess a demonstrated prowess in basic space technology which in the context of cold war rivalries carries with it an implied danger of military applications. This implied threat casts a shadow on the international political scene—as, of course, does our own military power and progress in space—especially since, with the present relative military standoff, space offers some chance of achieving a decisive military advantage over one's opponent. Insofar as either side convinces itself of the possibility of scoring a strategic gain by extending military competition into space, it will be under pressure to do so.

---

[1] The author is indebted to F. C. Ikle, A. H. Katz and R. S. Leghorn for discussions or comments on certain aspects of this paper, and especially indebted to Faith Wright for considerable assistance in its preparation.

Both the character and extent of military competition in outer space, therefore, are likely to depend considerably on the nature and intensity of the cold war, and in particular on what progress may be made in controlling its military aspects. If the cold war remains at its present level or even intensifies, it is possible that the major powers may find it in their mutual interest to declare some kind of truce in military competition in at least some areas. Military activity in space has not yet advanced far; and it seems likely to prove expensive and dangerous (from the standpoints both of effective command and control and of excessively large weapon yields) to a degree exceeding the predictable military advantage. Moreover, effective control arrangements seem feasible as indicated by present technology. These facts suggest that outer space may be among the most likely areas for effecting any arms restraints in the near future. However, even if such restraint should be exercised, the potentiality of military developments in space will remain as long as the basic military competition continues, and as such cannot be completely divorced from the cold war.

It must, then, be recognized that political factors constitute as much a part of the present setting of developments in space technology as do directly military considerations. This paper, however, is concentrating on the military aspects, and will begin with a review of contemporary military doctrine, principally as it concerns strategic forces, which will bear on military developments in space.

CURRENT MILITARY DOCTRINE

The dominant notion in current military thinking is that of deterrence. In military parlance this term refers to the effort to deter war by threatening a potential aggressor with retaliation or other consequences on a scale that would inflict greater damage on him than his expected gains would justify. The deterrence concept is applied to planning for military eventualities of every order. But as the primary concern of military planners today is to prevent big wars, the major emphasis is on deterrence by strategic nuclear forces.

The primary deterrent role envisaged for the strategic nuclear forces of the United States—principally the Strategic Air Command—is to maintain before a potential enemy a constant and convincing threat of nuclear retaliation of unacceptable proportions in the event the enemy were to launch a nuclear attack on the United States or one of its allies. It was formerly held (e.g., by John Foster Dulles) that retaliation by the strategic nuclear forces might also serve to deter other, nonnuclear forms of aggression. However, as Soviet capabilities have grown, the possibility of retaliating with a major nuclear attack on the homeland of the Soviet Union or its allies in response to a nonnuclear agression has become increasingly remote. (There remains, however, a distinct possibility that the strategic nuclear forces might become involved in the expansion of a limited war in which tactical nuclear weapons were used within the battle zone.)

The fact that our strategic forces are primarily designed for use in response to a nuclear attack—for a "second-strike" role, as opposed to launching the initial strike—has important consequences for the characteristics of the forces. In particular, it imposes a requirement that the forces should be able to endure an attack undertaken by the enemy specifically to wipe them out, and still retain sufficient effectiveness to inflict overwhelming damage on the aggressor. This means that we must reduce to a minimum the vulnerability of our strategic forces to whatever forms of surprise attack a potential enemy might mount.

Such relative invulnerability can be achieved by a number of techniques used separately or in combination. These include a greater number of well-dispersed delivery vehicles (missiles and bombers) than the enemy has missiles or other vehicles with which to attack them; increased mobility (bombers on airborne alert, Polaris submarines, or rail-mobile missiles) and other techniques to improve concealment of weapons; radar or other systems to warn that an attack is under way and to provide time for further dispersal or concealment, such as putting additional bombers in the air; and heavily protected ("hardened") underground missile-launching sites which would require an enemy to fire several missiles at each site in order to be confident of destroying it.

The degree to which such relative invulnerability is achieved can have a marked effect on the likelihood of a general war. If a nation's strategic forces consist entirely or chiefly of highly vulnerable weapons, such as concentrations of bombers on a relatively few airfields or collections of unprotected missiles at a relatively few missile-launching sites, an enemy has an overwhelming incentive to attack such forces if he has any reason to suspect that they may be about to be brought into play. In these circumstances, the attacker could destroy several weapons for each one of his own expended and thereby rapidly achieve a position of decisive superiority. If both sides have highly vulnerable forces, the potentialities for trouble are even greater than if only one side does, for in this case each side will know that the other is under pressure to "go first," and this will itself increase the nervousness of the first side. A high degree of vulnerability also requires a nation to respond rapidly to any threat; it must be able to react on very short notice and launch its strategic forces before they can be destroyed on the ground.

On the other hand, if both sides possess a sufficient degree of invulnerability to be able to ensure heavy retaliation even if attacked first, the temptation to go first would be lessened. The pressures to initiate general war in response to equivocal evidence or warning would be much reduced since either side could afford to "wait and see" in the event one or a few weapons went off unexpectedly.

From the point of view of inhibiting general war, the best situation obtains when each side has forces so invulnerable that several enemy weapons would normally be required to destroy a single weapon. If this situation is accompanied by an approximate equality of strategic forces, then—if deterrence should fail and war break out—the strategic advantage would reside in going *second,* not in going first. To achieve this situation would not by any means be to eliminate all possibilities of general war; it could, however, be expected to eliminate the possibility of war initiated by calculated surprise attack undertaken specifically to eliminate the strategic forces of the opposing side. If each side is convinced that, if war does come, it could achieve a decisive superiority by letting the other side expend its weapons in unprofitable attacks against its own strategic forces, each side would be strongly motivated to wait and let the other start it, in which case it might never start.

For the purpose of achieving a high degree of relative invulnerability, some of the techniques mentioned above are better than others. In particular, systems that rely on short-term warning and require very fast responses to achieve additional dispersal or concealment (such as a 15-minute ground alert of a bomber force) seem more accident-prone, and therefore less desirable, than do weapon systems in which the weapons are permanently concealed or otherwise protected.

The characteristics of the weapons themselves are also highly relevant to this problem. Weapons of greater accuracy and higher yield are more useful for attacking protected strategic weapons on the other side than are weapons of lower accuracy and smaller yield. This accounts for the fact that weapons of this latter sort—weapons that are better for attacking cities and civilian populations than for attacking strategic weapons—may be less likely to provoke excessive nervousness on the part of the strategic planners and decision-makers of the other side. Although it is not possible in most cases to decide unequivocally the uses to which a given weapon may be put, it will be convenient to refer to weapons that are poorly suited to attacking strategic forces but well suited to attacking cities as "countervalue" weapons. Weapons that are especially suited to attacking strategic forces will be referred to as "counterforce" weapons; these may of course also be used for civil retaliation, but they will not be termed "countervalue" weapons in the sense used here.

Another factor bearing on the relative invulnerability of the strategic forces is the presence or absence of active defense systems of substantial effectiveness. (By "active" defense systems are meant those devised to actively intercept or counter attacking weapons, in contrast to "passive" defense systems involving no active response, e.g., hardening of missile sites.) It is not necessarily true that an active defense system is "defensive" in the classical sense; if an active defense system is used by an aggressor to nullify the retaliation of his victim, it would be highly "offensive" in character and would damage the strategic forces of the victim just as effectively as if they had been destroyed on the ground.

Whether an active defense system increases or decreases the likelihood of general war depends therefore very much on the character of the system. If such a system is good at protecting "point" targets such as missiles but not especially good for the protection of cities, it would simply enhance the invulnerability of the forces of the side possessing it without precluding possible countervalue retaliation by the opponent. On the other hand, if such a system is good at protecting cities, it will serve at least partially to preclude retaliation by the opponent; in this sense it has the effect of making the opponent's forces more vulnerable by decreasing their countervalue utility. If both sides possess active defense systems that are good at protecting cities as well as protecting strategic forces, their net effect will depend on their detailed functioning and characteristics, principally whether they are equally effective when they are alerted and when they are not. If, for example, they are far less effective if taken by surprise, they may be of much more use to an aggressor than to his potential victim.

In addition, the effective vulnerability of strategic systems can be influenced by the presence or absence of certain kinds of information-gathering systems, such as the *Samos* reconnaissance satellite, a point which will be discussed more fully below.

It is worth pointing out that, at least at the moment of writing, there is no consensus among students of these matters in the United States as to the detailed characteristics of the weapon systems we should adopt. There is of course a very general agreement that our weapon systems should be as invulnerable as seems economically reasonable. There is much less agreement as to whether we should adopt systems that are primarily retaliatory in a countervalue sense, or whether we should maintain a large component at least of general-purpose forces that have some counterforce utility, i.e., forces that have the effect of making the strategic forces of the other side somewhat vulnerable. At the present time we are maintaining both types of forces, though the bulk of our capability is in forces of the latter, counterforce variety. Our manned bombers, which can deliver very-large-yield weapons with great precision, and which can find targets whose precise location is not known in advance, are especially good for attacking strategic forces, while Polaris missiles, having much smaller yields and relatively poor accuracy, are more nearly countervalue in character.

The basic dilemma is as follows. To continue to maintain a considerable capability to attack the strategic forces of the other side has certain marked drawbacks: it tends to require a substantial superiority in strategic forces, it is likely to seem provocative to the other side, it is likely to continue or even heighten the arms race, and it may ultimately make general war itself more likely. The adoption of "purely" countervalue systems would be less likely to produce these effects. On the other hand, if general war were actually to break out, we should obviously wish to minimize the damage to ourselves, and this would be better accomplished for the near future by maintaining the best possible capability of destroying the strategic forces on the other side. (The qualification "for the near future" is necessary because the continuation of such a capability might over the long run lead to the development of very dangerous systems—systems capable of killing far greater numbers of persons, regardless of anything we might do about it, than is true today.) This dilemma has not been thoroughly resolved within the United States, a fact that accounts for much of the uncertainty surrounding the likely evolution of space weapon systems.

There are many refined considerations it has not been possible to include in the foregoing brief summary of current military doctrine. However, the discussion should serve as a sufficient framework for evaluating the possible military consequences of future space-weapon developments and arms control arrangements. As will be seen, these military criteria are by no means the only relevant considerations, but they form a vitally important part of the whole.

Possible Space-Weapon Developments

ORBITAL BOMBS

The most frequently discussed possibility of a new form of a space-weapon system is that of bombs in orbit. As usually envisaged, this refers to the storage of nuclear weapons in satellites in orbit about the earth at relatively low altitudes such as 150 to 500 miles, although they could conceivably be placed higher. In terms of economy, command and control, and accuracy such a system offers no advantage over storing the same weapons in ballistic missiles on the ground, to be launched only when the weapons were to be fired. This point has so far proved the dominant consideration; so far as I am aware, neither side presently has orbital bombs under development.

The argument usually advanced in support of orbital bombs, and one which is not without considerable military persuasiveness, is that the deployment of such weapons would provide an added degree of invulnerability to the total strategic forces. This is accounted for by two factors. First there is the fact that orbiting weapons can be protected by the use of highly effective concealment techniques and decoys; these will be discussed in a later section. Second, as additional weapons to be attacked, thus requiring different attack-weapon systems, they would complicate the task of a potential aggressor contemplating a surprise attack; the problem of coordinating an attack on a large number of orbital weapons with a simultaneous attack on ground- or sea-based weapon systems appears quite formidable.

If a reasonable number of orbital bombs were deployed in well-separated orbits, it would require a minimum of several hours after launching ground-based interceptors from a single country before the orbital weapons could all be destroyed, and the launching of the ground-based interceptors could probably be detected by radar or other means. This would provide ample warning time to put the ground-based bomber systems on airborne alert and to take other precautions of a similar character. (There is a possibility that a potential enemy might be able to deploy interception devices in orbit gradually and if undetected in the process, to execute a coordinated surprise strike at some later time; however, this possibility could be made very difficult to realize.) A potential aggressor mounting an attack on the ground-based components of the strategic forces before launching an attack on orbital weapons would face the prospect of prompt retaliation from the orbital weapons however successful the attack on the ground-based strategic systems might be.

The potential characteristics and utilization of orbital weapons cover a considerable range of possibilities. They could be used as a purely retaliatory system aimed chiefly or entirely at cities, in which case they could have relatively modest yields (in the region of 1 megaton). The problems of effecting re-entry of these devices from orbit with sufficient accuracy for this purpose can probably be solved without the necessity for some form of terminal guidance.

Another possibility of a purely countervalue system that has been discussed, and which appears much more disturbing, would involve placing in orbit a limited number of devices of very large yield (a few hundred megatons or more) which would be detonated at orbital altitudes (say 150 miles) rather than be brought down to earth before detonation. The thermal effects from such a high-yield device could set fire to a large fraction of a continent, the extent of which would probably be limited only to that which could be "seen" from the point at which the device was detonated, except that areas protected by cloud cover at the time of detonation probably would not be ignited.

In addition to purely countervalue possibilities, it may prove possible to deploy orbital-bomb systems that are effective for attacking the strategic forces of an opponent. This would probably require devices of moderately large yield that could be brought down out of orbit with considerable precision, possibly using some form of active guidance (such as television) in the terminal phase of delivering the weapon to its target. Systems of this type would probably need to incorporate, or be supplemented by, reconnaissance systems for gathering suitable target information. Also, if a system of this type were to be used for initiating a coordinated surprise attack, it would probably be necessary to "bunch" the weapons in orbit in order to effect the delivery on their targets within a relatively brief interval of time. Retaliatory weapons, on the other hand, could be spread out in separated orbits since it would not be necessary to deliver them all at nearly the same time.

It is at least possible to envisage a "changeable" system of orbital weapons in which the bombs were initially dispersed in well-separated orbits, and which would hence appear retaliatory in character to an opponent observing the system, but in which the bomb-carrying satellites had a capability of maneuvering in orbit so as to "bunch" the weapons together for a coordinated surprise attack.

Providing sufficient reliability in the control of orbital weapons of any kind by the national leaders and the military services concerned poses a difficult technical problem. The problem may conveniently be separated into two parts: first, ensuring that the device will function as intended when it is required, including protection against the possibility of enemy jamming of the radio communication link to the device; and second, ensuring that the device will not be detonated in some manner not intended, including protection against possible enemy "capture" of the control of the device. Although these problems are more severe in the case of orbital weapons than in the case of certain conventional systems for a number of reasons, including the facts that the control would be based exclusively on radio communication and that there would be no personnel at the receiving end in direct control, it is likely that they could be adequately solved. One suggestion advanced that would serve to alleviate the control problem to a modest extent is that of keeping the orbital weapons and their launching boosters on the ground until a sufficiently intense crisis were at hand, at which time they would be launched, possibly to be recovered from orbit later if the crisis were otherwise resolved. However, this suggestion is not without serious difficulties of its own.

It is possible that, with the passage of time, the development of orbital bombs may seem more attractive to some of the smaller industrial powers than to the two present major nuclear powers. Thus far the development and operation of relatively invulnerable strategic weapon systems, such as *Polaris* submarines and airborne alert, have proved to be quite expensive, and only the two major powers have been able to afford them. The British *Blue Streak* was canceled precisely because the missile was going to be too vulnerable and it was not thought feasible to protect it adequately within available British resources. The further development of booster rocket technology by the major powers may make possible the deployment of relatively invulnerable orbital weapon systems by some of the secondary industrial powers. This is only a possibility, but it is interesting to note that it was implicitly predicted by Louis N. Ridenour in a brilliantly imaginative playlet, "Pilot Lights of the Apocalypse," written over 15 years ago projecting a future state of military affairs which the march of technology has shown a disturbing tendency to bring closer and closer.

My own view is that at best the potential military advantages of orbital weapons of all kinds, including most especially the large thermal devices, do not outweigh the political and other motives for wishing to keep them out of existence, considerations which will be discussed in more detail later.

LUNAR-BASED BOMBS

A different type of space-based retaliatory system that has been discussed, though less seriously, is one in which retaliatory missiles would be emplaced on the moon. Again, the principal motivation for contemplating the deployment of lunar-based weapons is that such basing would provide added invulnerability in the sense of reliably increased retaliatory capacity. Since the time required for earth-launched missiles to arrive at the moon to destroy retaliatory weapons based there would be a day or more, mounting an attack on lunar-based weapons would provide warning of the attack well before the lunar-based weapons could be destroyed; if, on the other hand, the ground-based retaliatory weapons were attacked first in a surprise attack, the lunar-based weapons would still be available for retaliation. It is sometimes argued that lunar-launched retaliatory weapons, being "visible" for a considerable time in flight, would be highly vulnerable to interception. However, given the level of technology at which lunar basing would be feasible, this argument is not forceful. A more telling criticism is that lunar-based weapons would require more energy than comparable weapons stored in free orbits at lunar distances. A lunar-based retaliatory system is not likely to be feasible as an engineering matter within the decade of the 1960's at least, and the need for additional invulnerability is unlikely to reach the level at which lunar-based weapon systems seem attractive. This problem is practically certain to be dominated by political considerations.

ORBITAL INTERCEPTORS

There has been considerable discussion[2] about the possibility of using orbital interceptors in an active defense system against intercontinental ballistic missiles. It is generally acknowledged that effective defense against ballistic missiles in their terminal phase, i.e., when warheads are near their intended targets, will be extremely difficult if not impossible. This stems primarily from the facts that the warheads, which cannot be

---

[2] See especially *Aviation Week,* October 31, and December 5, 1960; *The New York Times,* June 6, 1961.

easily destroyed even when going relatively slowly, are traveling with extremely high velocities during the terminal phase. Neither of these facts is true of a ballistic missile in its early or 'boost" phase, when the rocket booster is in the process of bringing the warhead up to final velocity; rocket engines and fuel tanks are relatively vulnerable objects and travel relatively slowly during the early phases of launching. This has led to several proposals to station in orbit missile-interception systems which would be able to attack enemy ballistic missiles during the boost phase and destroy them before they were well on their way.

Some systems of this type are reported to be under active study or development. The terminology in use for designating these systems is somewhat confused. The term "BAMBI," standing for "Ballistic Missile Boost Intercept," appears to be a generic term for these systems, though a specific proposed system may also bear this name; the same is true of "SPAD," which stands for "Satellite Protection for Area Defense." One specific system of the BAMBI type, designated a "Random Barrage System," would employ 20,000 to 100,000 interceptor units in so-called "random orbits" about the earth. In this system each interceptor within sight of an ICBM at the time of launching would attack the missile during its launch phase. The requirement for such large numbers of interceptors is presumably generated by the fact that only a relatively small fraction of the devices would be in orbit over a country at one time. (These numbers indicate very strongly that conventional nuclear warheads are not contemplated for these devices.) SPAD, as a somewhat different system of this general type, would employ 2000 to 3000 satellites in orbit at approximately 200 nautical miles' altitude, each of them equipped with one to six interceptor missiles that could attack ballistic missiles during their boost phase.

There are obviously many formidable technical problems to be solved before any active defense system of this type could become operational. The extent to which such problems have been solved or can, within a reasonable range of expenditure, be solved is not yet clear. (Cost estimates of $15 billion have been given for a system of unspecified efficacy.) However, a few basic principles are clear. If such a system is not to shoot down all rocket boosters at all times and over all countries, which would preclude not only "peaceful" launches but also the launching of replacement units necessary to maintain the system itself, then the system could at most be "turned on" only over certain countries and/or at certain times.

Thus such systems may involve associated warning and readiness problems that will seriously impair its value as a defensive system against surprise attack. That is, systems of this kind may be intrinsically better suited to use by an aggressor for precluding the retaliation of his victim than as a device to aid a potential victim in defending himself against a no-warning surprise attack. If this should prove to be the case, there may be merit in implementing arms control arrangements to prevent the deployment of such systems. In addition to their possibly increasing the likelihood of general war, the presence of 100,000 interceptor missiles in orbit above the earth would introduce substantial "bookkeeping" and other technical problems for various applications of space technology, including peaceful ones.

MANNED VEHICLES

If any military weapon systems at all are deployed in outer space, it is likely that manned space vehicles will have important roles to play, though the exact nature of these roles will depend considerably on what other kinds of weapon systems are deployed. Several manned space vehicles are under active study or development by the military services at the time of writing, of which the best known is a device called *"Dyna-Soar."* The military application of manned space craft that has most often been discussed is that of inspecting and possibly destroying hostile or unknown satellites. It is even possible to envisage manned space craft that could "capture" an unknown satellite and return it to the ground for inspection. This type of capability could be quite useful in a number of contexts, including some applications to arms control.

Another possible application would be to certain types of reconnaissance missions, either for simple reporting or for directing weapons against targets of opportunity which might unexpectedly be encountered, or against targets whose precise location was not previously known. In connection with armed reconnaissance, however, it should be noted that personnel in space are much more vulnerable to radiation effects from nuclear blasts than are personnel on the ground since the atmosphere provides considerable shielding of such radiation.[3] In this absence of considerable shielding a man in space might be vulnerable to medium-yield thermonuclear weapons at perhaps hundreds of miles. The possible strategic significance of manned military space craft is difficult to assess, in no small part because the military environment in which they would operate is itself so uncertain.

SURVEILLANCE AND RECONNAISSANCE DEVICES

At least two types of surveillance and reconnaissance systems based in space are possible. These can be illustrated by systems currently under active development in the United States. The first is the *Midas* system, which employs orbiting satelites equipped with infrared sensors to provide warning of the launching of a ballistic missile attack by large rocket vehicles. Except to the extent that such a system might be used as part of an active missile defense, it would not appear to be of value in initiating an attack on enemy strategic forces; that is, its sole directly military effect

---

[3] RAND Corporation, *Space Handbook.*

would be to reduce the vulnerability of some components of the strategic forces by early warning. Since no data on the expected performance characteristics of the *Midas* system—i.e., detection sensitivity of the infrared detectors, the size of rockets that could be detected, the frequency of false reports of launchings ("false alarms"), and so on—have been released, it is difficult to estimate its effectiveness. It is clear that, as a matter of engineering and physics, a system of this type of almost any degree of effectiveness could be developed; but, depending on the number of satellites required and the frequency with which these complicated and expensive devices might need to be replaced, the economic feasibility of such a system may be another question altogether.

The second space-based surveillance system under active development in the United States at the present time is the *Samos,* which is intended to perform photographic reconnaissance. Return of the photographic information could be accomplished either by a television link or by direct physical recovery of the film package. Since direct film recovery may be expected to permit considerably better resolution than that achievable with television relay techniques, I shall discuss the case of recoverable-film satellites, on which considerable information has been prepared by Amrom H. Katz.

The performance characteristics of photographic satellites are usually discussed in terms of three characteristics: film resolution, focal length, and orbit altitude. Film resolution, which depends on the quality of the film and on the size and quality of the lens, is usually expressed as the maximum number of decipherable lines per millimeter of film. The best currently achievable film resolution appears to be in the region of 100 lines per millimeter, or .01 millimeters between adjacent lines. This requires large-aperture, high-quality lenses and special film, and is no ordinary accomplishment; typical high-quality photographs in normal use are more nearly in the range of 10 to 40 lines per millimeter.

The corresponding ground resolution—the minimum distance between two barely discernible points on the ground—is obtained by multiplying the minimum discernible distance on the film by the ratio of altitude to focal length. Thus, to achieve finely detailed ground resolution, it is desirable to have high film resolution, a low satellite altitude, and a long focal length. For example, a focal length of 36 inches and a film resolution of 40 lines per millimeter will yield a ground resolution of 20 feet from a satellite in orbit at 150 miles' altitude. In order to be easily identifiable in a photograph, a familiar object that is roughly round or square in outline must have a diameter several (five or more) times greater than the minimum ground resolution.[4] Long but thin objects, such as railroads,

---

[4] It should be stressed that ground resolution is only one of many factors affecting identifiability; others, such as illumination and character of the film emulsion, are of considerable importance.

need not satisfy this requirement in their narrow dimension. Of course, if the object is not familiar, it may be that no amount of resolution would be sufficient to identify it.

A possible range of ground resolution achievable from satellites at 150 miles altitude has been given by Katz. These values range from 60-foot ground resolution, with a 12-inch focal length and 40 lines per millimeter, to 2.4-foot resolution, with 120-inch focal length and 100 lines per millimeter. The accomplishment of 2-foot ground resolution from satellite-borne cameras would be a distinctly impressive achievement, and values of 8 to 10 feet or more are perhaps more realistic, at least for the next few years. For comparison, mapping photography from 30,000 feet (6 miles) during World War II yielded a ground resolution of perhaps 15 to 20 feet. It should be noted that the achievement of anything like this degree of ground resolution with techniques capable of penetrating darkness and/or cloud cover, such as radar or infrared mapping does not seem to be a likely development.

Let us now consider the possible utilization and strategic significance of photographic reconnaissance satellites. The possible technical capabilities of these devices have been given in some detail because, while it is true that they may yield important information, they also possess limitations and liabilities which should be understood. Discussions of the *Samos* system in the United States almost always concentrate on the maximum possible gains from the system, and rarely include consideration of possible limitations or liabilities.

In the interests of limiting the arms race it is important for each of the major powers to have reasonably accurate inventory information on the strategic force levels of the other. If either side has exaggerated estimates of the other's capabilities, or has estimates thought to be of insufficient reliability, it will tend to build strategic systems in proportion to the worst possible threat, and this in turn is likely to provoke the other side into building up its strategic forces to the originally suspected level, or beyond. (Something of this kind seems actually to have happened in connection with our own strategic bomber program.)

The Soviets do not seem to have assimilated this point, though it has been made to them repeatedly. At the present time, they seem quite hostile to the operation of the *Samos* system, although it probably is in their interest that we should not overestimate their strategic force levels or be nervous about uncertain possibilities.[5] (There is a possibility that the Soviets would not think it in their interest for us to have good inventory information if their forces were quite weak; exaggerated estimates of their strength by the West might then be seen as operating to their advantage.)

---

[5] Richard S. Leghorn, "The Pursuit of Rational World Security Arrangements," in *Arms Control, Disarmament, and National Security,* George Braziller, Inc., New York, 1961.

One understandable reason for Soviet nervousness over our acquisition of such information is that, with conventional fixed missile sites and bomber bases, it is difficult to obtain reliable inventory information about force levels (i.e., numbers of weapons and carriers) without also obtaining targeting information (i.e., locations of weapons and carriers). Targeting information would have the effect of making the forces under observation considerably more vulnerable than if they were successfully concealed—a problem of particular concern to the Soviets who, by virtue of their more closed society, enjoy an advantage in the current asymmetry of targeting information. This has led some Americans to suggest that the Soviets should adopt mobile weapon systems which would lend themselves to being "counted" by a *Samos* system, but which, in view of the time required for return of the film package, could not be pinpointed by the system for targeting purposes. Of course, some Americans are in fact interested in *Samos* for obtaining target information, in the event a general war were to arise. However, the more often expressed motivation for the development of *Samos* is that of providing inventory information on force levels; in particular, this is the argument advanced by those who advocate its use in arms control arrangements and the stabilization of the arms race.

It is usually assumed in these arguments that *Samos* would provide this information, or contribute substantially to providing it with sufficient reliability. There is no doubt whatever that the operation of *Samos* would yield considerable and useful information. It is also possible, though far less certain, that it would yield the desired inventory information on force levels—provided that the Soviets build fixed missile sites that can be recognized as such, and/or build mobile systems that can be recognized and "counted," and that they do not build things which look like missiles or carriers but which in fact are not.

These are critical provisos, however, which deserve examination in further detail. It seems to me that, with the aid of periods of darkness, occasional periods of cloud cover,[6] and only a modest amount of ingenuity, it would be possible to build fixed missile sites one after the other ("like sausages," as Khrushchev has put it) that would not be identified as such by an eight-foot ground resolution capability in the reconnaissance system. It also seems highly likely that a nation could construct and deploy train-mobile systems for the smaller, solid-fueled missiles, such as *Minuteman,* so that they could not be identified as mobile missiles by a system of this resolution capability. And it is virtually certain that one could build relatively inexpensive objects that looked precisely like conventional missile sites on photographs with eight-foot ground resolution but which in fact were decoys. These same statements are likely to remain true even if the

---

[*] There are large areas of the Soviet Union that are covered with solid clouds for weeks at a time.

eight-foot resolution were replaced by a two-foot resolution, though with somewhat less forcefulness.

The possibility of building fixed or mobile missile launching facilities that would not be detectable by *Samos* is basically an economic question. It is certainly true that it would be more difficult and expensive, other things being equal, to construct a missile site in such a way that the construction activity could not be detected and identified by *Samos*. However, other things may well not be equal. In the missile sites we have most recently been building in the United States we have been devoting an average of several million dollars per missile to the construction of heavily shielded underground launch sites which would protect the missile against blast and ground shock from relatively close nuclear explosions. If the missiles were instead sufficiently well concealed (an alternative presumably more feasible for the Soviet Union than for the United States), there would be much less motivation for the heavy protective construction, and the cost of building the launch facility in a sufficiently concealed manner might well be less than that of providing protection through sheer concrete. Even if the protective construction is included, it does not seem likely that the incremental cost of hiding the construction activity would be more than about 25 per cent of the cost of a similar but unconcealed site.

It thus seems reasonable to conclude that there is no certainty whether the *Samos* system would provide reliable inventory information, essentially no inventory information, or, possibly, misleading information which might indicate that force levels were either higher or lower than the actual case. This evaluation, which should be regarded as tentative, is relevant to the extent to which we might be willing to abstain from deploying the *Samos* system if there were some other advantages to be obtained in not doing so.

DECOYS AND CONCEALMENT

It is desirable to include some discussion here of a few supplementary techniques which, while not weapon systems in their own right, might well be employed in conjunction with some of the possible space-weapon developments sketched above.

If a military satellite, such as an orbital bomb or a reconnaissance satellite, is likely to be attacked by a potential enemy, it can be protected partially or completely by any of at least three techniques used singly or in combination. The first of these is the *use of absorptive coatings* on the satellites, or special configurations of absorptive material, to render the satellite difficult or impossible to detect by ground-based radar. Coatings are available that can reduce the radar power reflected from a target by a factor of 1000. It seems probable by using a combination of special satellite shapes and such absorptive material, militarily significant satellites could be made to have an apparent size as measured by radar, known as the "radar cross section," of one square centimeter or less. Radar systems that could detect such a target in an orbit of a few hundred miles' altitude are not likely to be achieved in the foreseeable future. (It should be noted, however, that it would be very difficult or impossible to similarly conceal an entire rocket vehicle during its launch phase.)

A second technique that might help protect orbiting military devices is the *use of decoys*. With light materials it might well prove possible to deploy dozens or even hundreds of decoys in orbit for each significant military device orbited, in such a way that the decoys would present the same appearance to ground-based radar or infrared sensors as the significant device. Ground-based pilotless satellite interception missiles will be very expensive machines, and the prospect of expending, say, 200 interceptors in order to destroy one genuine target would appear very discouraging. One of the motives for employing manned spacecraft for satellite interception missions is that human pilots could be expected to distinguish more readily between decoys and relevant targets.

It is worth pointing out that neither of these techniques, alone or in combination, may be sufficient to defeat a combination of several detection techniques, including radar, infrared, and optical techniques used together.

The third technique worth discussion here is the *use of variable orbits*. Since the orbit parameters of a satellite can be determined with increasing accuracy as the number of revolutions through which it is observed increases, it is relatively simple to program the trajectory of an interception missile against a satellite that has been in orbit for some time but much less so when the satellite must be destroyed on its first or second pass. It is possible to envisage the use of satellites that retain sufficient thrust capability to be able to change their orbits in ways that could not be predicted by a potential enemy. This would make the interception mission much more difficult, though not necessarily impossible, especially if manned spacecraft that could "chase" an evading satellite are used. (This technique would not of itself make detection difficult, in contrast to the two preceding techniques.)

In contrast to protecting space-based systems, it is possible also to harass them by techniques short of destruction. Satellite communication systems, or the communication links on military devices in space, can be jammed or "spoofed" with false signals. Infrared or other sensors, such as those used in the *Midas* system, can also presumably be "spoofed" with false signals which would simulate targets of the type the system was intended to see. Whether any of the major powers would find it advantageous to engage in such provocative and possibly dangerous behavior is, however, far from clear.

POSSIBLE STRATEGIC DEVELOPMENTS

The foregoing catalog of military techniques and systems proposed to date for outer space does not in itself provide any indication of what the military environment in space might actually look like in, say, 1970 or 1975. This will depend very much on the evolution of events and international tensions over the intervening years. Even on the assumption that no major general war intervenes, the possibilities range from complete military restraint, possibly extending to the internationalization of space, at one end, to active forms of limited space warfare at the other. To provide some sense of the possible range of circumstances, it may be useful to sketch in very rough outline three possible types of strategic military environments as they might exist in the late 1960's or in the 1970's.

The first possibility is an environment characterized by relatively *strenuous military competition in space,* perhaps extending to limited forms of warfare in space. There would be emphasis on the active development of systems designed to attack strategic forces of possible opponents, including intensive use of reconnaissance systems and sensory devices of various kinds. Each side might well feel moved to attack the reconnaissance systems and orbital weapons of the other side whenever it could locate and identify them. This situation would probably lead to a race in the development of detection, tracking, and identification techniques on the one hand, and concealment, evasion, and decoy techniques on the other.

Some people argue that the advent of limited outer space warfare would be a healthy development since it would furnish an outlet for military competition that would not itself be destructive of life and would not be likely to spread to the earth proper. Others, however, argue that warfare in outer space would be distinctly dangerous because it would seem relatively painless to initiate but would thereafter have a high likelihood of spreading to a more destructive form of conflict on earth. It does not appear to me that it is wise to attempt conjectures on the relative merits of these two views.

The second type of illustrative strategic environment that might evolve is one in which the major powers would deploy *weapon systems purely retaliatory in character,* with outer space otherwise employed primarily for peaceful purposes. The retaliatory weapons might consist of large thermal devices to be detonated in orbit or of orbital bombs of intermediate yield that had to be brought down to their targets to be reasonably effective. Space-based reconnaissance systems might be employed, perhaps cooperatively or under an international agency such as the United Nations, to assist in maintaining inventory information on strategic force levels but not with the objective of providing targeting information. (This is technically feasible if mobile weapons are employed.)

The third illustrative environment would be one in which *considerable* (*possibly complete*) *restraint* might be exercised in the deployment of national weapon systems in outer space. Possibilities in this category range from informal or even tacit restraints of the "I-won't-so-long-as-he-doesn't" variety up to the complete internationalization of all activities in outer space. Some of these possibilities will be discussed in more detail in the next section.

It is worth stressing again that these are purely hypothetical projections. There is at present no clear basis for assuming that one type of environment is much more likely to emerge than another, or that some other, still different type of environment may not evolve.

Possible Control Arrangements

We have noted that the major powers may exercise self-restraint in the militarization of space. It might take the form of simply waiting to see what the other side is going to do and not doing many things that he doesn't, in which case the deployment of military devices in space might not progress very far beyond observation satellites. Such restraint might also entail informally negotiated arrangements. For example, we might have our ambassador inform the Soviets privately that, at least until further notice, we would not deploy an active defense system of the "Random Bombardment" variety if they did not; we would then rely simply on unilateral intelligence and surveillance to see whether or not they did deploy such a system. (An undertaking of the immensity of a Random Bombardment System would be most unlikely to escape notice.)

There is, however, much potential merit in providing substantial reinforcement of such restraints by means of formal, negotiated arrangements; the balance of this section will be devoted chiefly to a rough outline of some of the possible negotiated constraints that have been proposed, making no pretense to exhaustiveness in the list of possibilities.

INTERNATIONALIZATION OF SPACE

In the eyes of most students of these matters, the most attractive solution to the problem of controlling weapons in space would be the complete internationalization of all activities in outer space. Placing all outer space development, operation, and exploration in the hands of an international agency would not only end military competition in space; it also would eliminate other, nonmilitary competition in space. Such an agency would presumably have no interest in the development and deployment of orbital weapons of mass destruction. Placing all space development under international auspices would also make it possible to evaluate such projects as putting a man on the moon more nearly in terms of the scientific merits of the project rather than in terms of national political considerations. However, given the present prestige attached to individual nations' efforts in space competition as an aspect of the cold war, the prospects of persuading the major powers to relinquish their national space programs do not seem auspicious.

Similar misgivings apply to the proposal offered from time to time for the complete prohibition of all outer space activity. There was a time when this seemed both desirable and potentially feasible, but it is now doubtful that a complete prohibition either could or should be obtained.

There may be somewhat better prospects for the partial internationalization of space programs—that is, the transfer of certain activities to an international authority, while retaining others under national control. It may prove possible to achieve agreement on international agencies that would be responsible for the development and operation of reconnaissance and warning systems of the *Midas* and *Samos* varieties in the military sphere, and of selected peaceful applications such as meteorological survey satellites, communication satellites, navigation systems, and so on. Such partial internationalization would probably, though not necessarily, be accompanied by some selective constraints on national space programs of the type discussed below.

Among students of arms controls, the system most often discussed as a candidate for transfer to international control is the *Samos* photographic reconnaissance system. The attractiveness of this possibility stems from the potential usefulness of *Samos* as an inspection technique in connection with other possible arms control arrangements. While, as has been noted, the reliability of this system for providing inventory information on strategic force levels when used alone may be open to question, there is no question about its usefulness as a supplementary technique, especially if aerial reconnaissance from manned aircraft were ruled out.

As an alternative to international operation and control, it is of course possible that the results obtained from *Samos* might be made available through the United Nations to all countries, to be tied perhaps to international arms control arrangements, while the United States (and other countries) were free to continue their unilateral efforts. It is also conceivable that a bilateral scheme for operation of *Samos* jointly by the United States and Soviet Union might be devised. (On the other hand, one argument sometimes raised against international operation of *Samos* is that since it would make the operating results of the system available to all the participants, weaknesses of the system would become precisely known to a potential evader.)

It has also been suggested, that an orbital boost-phase interception system of the BAMBI or "Random Bombardment" variety might be operated under international auspices as an arms control measure intended to inhibit the possibility of a surprise missile attack by a potential aggressor. Assuming that the necessary degree of cooperation could be obtained to begin with, there seems little reason to spend $15 billion for a system whose net effect at best would be to enforce constraints that would seem to be more easily implemented on the ground. Moreover, the deployment of a Randon Bombardment System would clutter up outer space immensely, creating problems of collision hazards, occlusion of radio and optical stars, and enormous "bookkeeping" requirements to keep track of which object is which. In addition, the system might well prove much more effective in precluding retaliation by a victim than in precluding the attack of an aggressor. Still, the possibility should perhaps be mentioned.

CONSTRAINTS ON SPACE PROGRAMS

There is a considerable array of selective constraints on national space programs which would fall well short of internationalization but some of which might be highly desirable. Five possibilities are discussed here.

On Orbital Weapons of Mass Destruction—A first example, and one that has been proposed by the United States in international negotiations several times since March 1960, involves the prohibition of weapons of mass destruction in orbit. In view of the fact that orbital retaliatory bombs might contribute an additional type of relatively invulnerable strategic weapon system and thereby reduce even further the incentive for a potential aggressor to launch a surprise attack, some students of these matters are of the opinion that orbital weapons should not be prohibited. My own view, however, is that it is likely to be in the greater interests of both national and world security at the present time to prohibit these devices, at least if certain surveillance and monitoring problems can be solved in a reasonably acceptable way.

This view is based on the following judgments, some or all of which may be open to question. a) The achievement of sufficiently secure strategic weapon systems seems entirely feasible without the necessity of adding orbital weapons. b) Medium-yield orbital bombs that would have to be brought down to their targets to be effective are likely to be somewhat vulnerable to active defense systems of the Nike-Zeus type.[7] The type of orbital bomb likely to seem most attractive to certain military planners would therefore be the high-yield thermal devices that could be detonated in orbit. We should most emphatically not build and deploy weapon systems in which a single accident could destroy perhaps half a continent. c) As Herman Kahn points out, the technological arms race is dangerous in its own right, perhaps most especially as it makes increasingly dangerous technology available or potentially available to an increasing number of nations, and substantial efforts to curb this race are thoroughly justified.

---

[7] Nike-Zeus is a terminal-phase warhead interception missile system under current development.

d) I would not attempt to predict domestic reaction within the Soviet Union to the presence of American bombs in orbit, but I believe that internal American reaction (particularly in the Congress) to the presence of Soviet orbital bombs overhead, most especially bombs of the large thermal variety, would quite possibly be of a character that would have distinctly adverse national and international effects. e) As with other possible measures of arms control, the prohibition of orbital weapons might contribute, however slightly, to an increased degree of cooperation in regulating armaments, and the habit of arms cooperation with potential enemies in areas of common interest is one well worth acquiring. To my mind, these considerations very clearly outweigh the possible improvement in relative invulnerability that might be achieved by the deployment of orbital nuclear bombs.

On Orbital Weapons in General—Instead of prohibiting only orbital weapons of mass destruction, one might prohibit all orbital weapons. With respect to weapon systems that have been discussed up to the present time, the chief impact of this added degree of generality would be to prohibit the deployment of anti-missile defense systems employing orbital interceptors of the "Random Bombardment" variety. In view of the unattractive features of such systems, as discussed earlier, there is much merit in prohibiting these systems as well. Indeed, one can probably make a stronger military case against orbiting boost-phase interception systems of this type than against orbiting weapons of mass destruction.

On All Military Systems in Space—It is possible to contemplate going yet further and prohibiting all military devices in space. As is the case in many prospective arms control measures, it would be quite difficult to draw a line that put "military" devices on one side and "nonmilitary" devices on the other. The phrase "all military devices" would, however, probably encompass warning and reconnaissance systems of the *Midas* and *Samos* varieties, respectively. If other things were equal, it would probably not be in our interest to prohibit such observation satellites. However, the Soviets may be more attracted to the possibility of prohibiting all military devices (their disarmament proposals of 2 June 1960 contained wording to this effect) than to the possibility of prohibiting all orbital weapons or all orbital weapons of mass destruction. Depending on the projected efficacy and ultility of the *Midas* and *Samos* systems, it may or may not seem reasonable to "trade" these systems for a prohibition of all orbital weapons. I would emphasize, however, that the apparent desirability of this "trade" would depend very strongly on the technical efficacy and utility of the observation satellites concerned, in addition to depending on the extent to which it seems important to prohibit orbital weapons.

On Military Development Efforts—It is at least possible to envisage selective constraints on the development of space technology that would be designed to inhibit the further development of military systems but would permit the continued exploration of outer space as well as the development of, perhaps, some new forms of space technology. It is, of course, not possible to draw an absolutely sharp line and decide that some technology is "military" and that some other technology is "peaceful," but it is certainly possible to identify some developments that would be of immediate importance to military applications and which might in addition be dangerous in terms of excessive devastation capacity or uncertainty of control. (The possibility of effecting selective developmental constraints of this type has sometimes been denied altogether, though not by people who have devoted any appreciable amount of thought to the subject.)

Possible constraints of this type include prohibition of the further development of boost-phase guidance systems, requiring all rocket firings after a given date to be implemented with existing boost-phase guidance; prohibition of the development of high-precision terminal guidance systems; prohibition of the further development of large solid-fuel boosters; and prohibition of all further development of large booster engines, requiring all firings after a given date to be implemented with existing rocket engine designs. Not all possible constraints of this type would be equally desirable, and considerable further study of the problems of implementation as well as the consequences would be required before any proposals of this type could be advanced with confidence.

It should be noted that implementation of constraints of this type would probably require some kind of technically qualified political body to pass judgment on whether or not a particular proposed development did or did not include prohibited development, and it would almost certainly require highly detailed inspection of the design and construction of space vehicles before launching. The current political feasibility of these measures, therefore, seems rather dubious.

On Outer-Space Weapon Tests—The prohibition of outer-space nuclear weapon tests represents another form of arms control in outer space, and one which has been much discussed at the Geneva Test Ban Conference and elsewhere. Despite the failure of this conference to reach any agreement by late 1961, there is a reasonable chance that nuclear weapon tests in outer space will be effectively prohibited by tacit agreement if not by treaty.

INSPECTION TECHNIQUES

Some brief indications of possible surveillance techniques for monitoring arms control arrangements of the type discussed above may be useful. An elementary but somewhat more detailed discussion can be found in the *Bulletin of Atomic Scientists* (May, 1959).

At least for the relatively near future, most of the possible constraints on weapons in outer space, such as a prohibition of orbital bombs, would probably require some form of global launch surveillance to monitor all launchings of space vehicles and/or surveillance of objects in near-earth orbits, coupled with some form of pre-launch ground inspection of devices that were to be launched. This dual arrangement of pre-launch registration and inspection of all authorized rocket firings and surveillance to detect large rocket firings or vehicles in orbit is the only form of monitoring envisaged for constraints on space weapons in the near future, and possibly the only practical and reliable method for the indefinite future. In this context it is usually envisaged that enforcement would be provided by destroying any detected satellites that had not been registered and subjected to pre-launch inspection.

It would be considerably easier to destroy such a clandestine satellite than to subject it to detailed inspection. Although it is conceivable that at some future time it may become possible to make detailed inspections of satellites in orbit with the aid of manned spacecraft, it is likely to remain a difficult and expensive operation at best. In any event, it will not be achievable for at least the next several years. But it is already possible to maintain a general inventory of satellites in orbit, as is currently being done; and, although such auditing techniques would not provide detailed information on individual satellites, it is likely that they could provide notice of any unauthorized satellites in orbit if sufficient resources could be devoted to the effort—a likelihood conditioned by satellite evasion and detection techniques already discussed.

With respect to the problem of monitoring all major rocket launchings, it is an unfortunate fact that such firings can be mounted anywhere in the world—from uninhabited islands in the oceans, from remote countries in Africa, and from the decks of ships, as well as from known launching ranges (such as Cape Canaveral) on the territory of the major powers. If submarines and rockets could be developed in such a way that significant payloads could be launched into orbit or outer space from submarines, this would complicate the problem even further, although at present writing it seems probable that any such submarines would be of a size that could hardly escape notice.

The fact that rockets can be launched anywhere in the world means that to render the launching of clandestine rocket vehicles absolutely impossible, global launch surveillance would be required. It presently appears feasible, as a matter of engineering and economics, to provide such global launch surveillance, although a global network providing very high reliability would be quite expensive, probably in the range of two to five billion dollars. Depending, however, on the degree of reliability required and the nature of the information demanded of the monitoring system, estimates ranging from $500 million to $10 billion can be obtained.

Several techniques are available to provide this type of surveillance, and they might be used singly or in combination. The most conventional and direct approach would be to employ a system of ground-based (or, as required, ocean-based) conventional radar sets. The number of installations required would depend on how soon after launching it would be desired to detect a rocket, and this in turn depends on the nature of the constraints being monitored and on such matters as how precisely it is desired to locate the point of launch of a rocket. If it is desired to detect essentially all major rocket firings (including ICBM tests as well as payloads fired into orbit), a minimum detection altitude of a few tens of miles would probably be appropriate, in which case the number of radar stations required can be roughly estimated as $16,000/h$ where h is the minimum detection altitude measured in miles. Thus to detect all rockets fired above 40 miles altitude by a system of this type, roughly 400 stations would be required, while to detect all firings above 80 miles altitude, approximately 200 stations would be sufficient. The average cost of individual stations of this type would probably be roughly 10 million dollars, where "roughly" is used to mean to within a factor of 5. The cost of individual stations would depart considerably from the average cost, depending on the local environment; arctic stations and floating stations would be considerably more expensive.

There are other possibilities besides conventional ground-based radar. It would be possible to employ a system of airborne infrared detectors in circulating aircraft which would be subject to essentially the same geometric constraints relating the number of stations with detection altitude as in the case of ground-based radars. It might also be possible to employ a system of orbiting satellites with infrared sensors similar to the *Midas* surveillance system, but the geometric and orbital relationships in this case would be considerably more complicated. Other potentially useful techniques include a different type of radar known as "high-frequency" or "ionospheric" radar and acoustic detection techniques.

In the event it is not required to detect the launching of essentially all large rocket vehicles, much more modest systems may suffice. For example, if it is only required to detect satellites in relatively low earth orbits, without precise indication of their points of launch, a very few radar installations —perhaps as few as one—might suffice. On the other hand, it is possible to envisage arms control arrangements of considerably greater complexity in which a system of global surveillance would need to be supplemented with high-precision tracking equipment on certain designated firing ranges.

INSPECTION REQUIREMENTS

At the present time we have little sense of the extent and reliability of inspection that should be demanded of any arms control arrangements. This is a very general problem which arises in the design of all or nearly all potential arms control arrangements, not merely those relating to outer space. The problem seems especially acute in this context at the present time because of currently active proposals to prohibit orbital weapons. The lack of what might be called an "inspection rationale" accounts in part for the enormous range of uncertainty in the cost of potential monitoring systems; but at the moment of writing, adequate system studies have not been performed on the basis of any criteria whatever.

It is quite often assumed in elementary discussions of arms control that inspection must be "foolproof." It is certainly true that political and strategic planning would be considerably simplified if "foolproof" monitoring were achievable, but there are many cases in which absolutely reliable inspection may not be achievable for political, economic, or basic technical reasons. This does not mean that arms control arrangements involving less-than-perfect monitoring are not desirable, but rather that the criteria for evaluating them are more complicated.

Many critics of imperfect arms monitoring arrangements overlook the fact that we continually employ the more complicated criteria in evaluating our unilateral defense establishment, which is far from perfect and has many "holes" which for one reason or another have gone unfilled. For example, it is quite well known that the United States has a warning system made up of a few radar installations in far northern latitudes (the "BMEWS" or "Ballistic Missile Early Warning System") which is designed to detect incoming enemy missiles fired over the north polar regions. However, it is also quite well known that the Soviets possess boosters fully capable of firing highly impressive warheads the long way around which presumably could not be detected by the BMEWS system. So far as is publicly known, there is no comparable system to detect enemy missiles incoming from southern latitudes. This does not mean that the BMEWS system is not useful, but only that the character of its utility is more complicated.[8]

These remarks should not be construed as an argument for "inadequate" inspection. They attempt only to point out that people who insist on "foolproof" inspection may be unconsciously guilty of employing a double standard. Some of the more relevant criteria for evaluating prospective arms control arrangements, especially criteria relating to military strategy, have been explored in a book by T. C. Schelling and M. H. Halperin, *Strategy and Arms Control.*

Conclusions

It will already be clear to the reader that no sharply drawn conclusions in the sense of predictions of the extent and rate of militarization of outer

---

[8] Whether it constitutes a billion dollars worth of utility is a matter of some debate among students of these matters.

space are in order. The feasible alternatives are too many, and the decisions yet to be made on central political and military considerations are of too great potential influence, to justify such predictions.

On the other hand, having reviewed what appear now to be the principal alternatives, I should like to emphasize my personal view that considerable efforts to prevent the extensive deployment of military weapons in space are fully warranted. This view stems in part from the exceptionally dangerous character of some of the weapons that may be deployed in space if the arms race is once allowed to move to that environment, but even more from a conviction that the technological arms race is itself highly dangerous and should be retarded in any reasonably secure ways that present themselves. The intrinsic hazards of the arms race derive partly from the generally increasing destructiveness of the systems, partly from their increasing complexity and the correlated difficulties of comprehending their operation, and partly from the fact that certain kinds of very destructive technology are likely to become increasingly cheap and within the reach of an increasing number of nations. Any readers who view the recent evolution of military technology with equanimity, or who otherwise think that all possible forms of arms control should be ignored, should read Ridenour and Herman Kahn.

It is not impossible that outer space, not yet having become an area of active military competition, may the more readily become an area of limited military cooperation among the major powers. Such cooperation, once established in outer space, might well prove "catching" in other areas of potential mutual interest in the limitation of armaments.

7.

The prospects for law and order

♦Lincoln P. Bloomfield

With each new successful space shot, whether by the United States or the Soviet Union, new contributions are made to man's potential for new knowledge, for far-ranging adventure, and for military power. But at the same time ever greater problems are accumulated for the statesmen of the world who face the task of somehow bringing space technology within a framework of international order before the problem becomes impossibly complicated.

Fifty-odd satellites have been successfully fired into the earth's orbit by the United States, and more than a dozen by the Soviet Union. More than half of the American satellites are still in earth orbit and almost half of these are still transmitting useful information. (Only one Russian satellite is still in orbit and has long since become silent.) In addition, both the United States and the Soviet Union have orbited "planetoids" around the sun, and Russia has made a hard landing on the moon, as well as photographing the far side of our natural satellite. Without the slightest doubt, a brand new situation is already in existence, with law and order, as usual, lagging well behind.

There is no immediate prospect of bringing outer space fully into the world's legal and political system. But there are increasingly good reasons why the effort should be made to improve that system so that it might cope effectively with the space age. In the meantime there are such seemingly small but politically complicated and difficult matters as the fixing of responsibility for damage done by fragments of falling satellites. The acute political complications which can result were hinted at on November 30, 1960 when the pieces of an American launching rocket carrying the Navy's *Transit* navigation satellite which was deliberately destroyed by the range safety officer at Cape Canaveral fell on an uninhabited area in eastern Cuba at a time when United States-Cuban relations were approaching an all-time low.

Another example of a practical problem which cannot await· a more rational world order is the cluttering of radio channels with signals to and from earth satellites on frequencies which impinge on those used for workaday activities on earth, a problem which will predictably intensify as communication satellites become operational and revolutionize—as they will—the sending of messages on earth. It will be enlightened self-interest rather than utopian dreaming that will bring order into the radio spectrum. Self-interest will probably make possible international cooperation in the science of weather forecasting, soon to be improved out of all recognition by the *Tiros* family of satellites.

But beyond all this lies the realm of military security, and it is here that the most profound issues of law and order will increasingly arise. Security here, no less than elsewhere, is a two-sided coin. One side of it is the urgent problem of keeping up with one's opponents in space so that no catastrophic imbalance can take place in the arms race; the entire philosophy of deterrence as a preventer of war depends on equilibrium, if not supremacy. The other side of the same coin is the urgent need for international agreements on control of outer space so that the already immeasurable dangers inherent in the arms race on earth will not be placed completely beyond human management by, for example, the stationing of nuclear weapons in orbit—a development with the most portentous implications for world security.

An array of formidable problems thus confronts man when he looks into space. But none of them can be resolved *in* space. They can be dealt with only on earth, in statesmanship, in carefully thought through military policies, in diplomatic negotiation, and in creative and imaginative planning in the fields of international law and international organizations. The purpose of this paper is to examine briefly the recent record of such thinking and action as has been taken in these two realms, and to suggest some of the relevant policy decisions which face the United States and, indeed, the world community as a whole.

The legal and political realms crisscross at virtually every point, and any separation between them must necessarily be artificial. The problems of law are in important ways the subject areas for international control; the problem of international control is in a sense the application of legal principles through international agreements. For convenience, however, this chapter examines the two fields—law and politics—separately. (I shall not cover two related problems—the area of international cooperation at the non-governmental level, which is treated in Dr. Odishaw's chapter, and the problems of arms control in outer space, which Donald Brennan deals with.)

The Law of Outer Space

The space age is in its infancy, dating from Octover 4, 1957—the flight of *Sputnik I*. In an earlier day a few hardy legal pioneers such as John Cobb Cooper and Andrew Haley applied their legal talents to the issues of outer space. But in the main, space law is a relatively new development; it was only in 1953 that the first doctoral dissertation on space law was prepared (by the Prince of Hanover at the University of Goettingen), an act which made the author's reputation. In general, the origins of space law are in the field of air law, itself a development of the 20th century. Behind both of course lie the principles of common law, reaching back into Roman times.

In the space age so far—the 1950's and early 1960's—the legal profession has taken the lead in studying the non-technical problems of outer space, specifically the legal issues. Indeed, it is likely that in terms of sheer volume of output, the lawyers have, if not exceeded, in any event run a close second to the rocketeers. Legal writings are proliferating as more concrete problems arise and more minds focus on the space problem. Yet despite the voluminous literature, the state of the art in space law is still relatively primitive. It is still being asked whether it is desirable to codify the existing rules or to allow them to develop case by case; whether outer space should be considered analogous to the high seas or perhaps to Antarctica; what the relationship is between air space above land, over which nations have absolute sovereignty, and outer space; where space begins, who owns the planets and celestial bodies, who should have jurisdiction over activities carried out in outer space, and what the rules of the road are, if any. There are no formal agreements on record about the activities that have been conducted so far and we are only making a beginning in anticipating activities only a few years off involving objects constructed in space, or on the moon or on the planets. There are problems concerning legal transactions which may take place in space; acquisition of territory; rights to natural resources; interference from space with rights on earth—a whole gamut of questions about which activities are permissible and which impermissible.

But if there is no agreement in the sense of a statute on the books, a treaty signed, or practice become embedded in custom, or even significant unanimity among the parties, there is a growing body of literature which in the fashion of international law can become the source for legal judgments. There are also some modest areas of consensus between the only two nations significantly involved so far. These will become apparent as we proceed to take up the specific issues of space law.

Perhaps the most basic question—and the most intriguing—is the matter of national sovereignty in space. Where are the boundaries of space? Where does outer space begin? How high up can nations exercise their traditional rights of sovereignty, and what are the rules above that height? Other questions may be more immediately practical; but few have been more thoroughly discussed and disputed.

THE PROBLEM OF SOVEREIGNTY

How far up does the nation's sovereignty extend? To answer this question an analogy from private law reaching back into Roman times has been frequently invoked: *Cujus est solum, ejus est usque ad coelum* (Who owns the land owns it up to the sky.) This maxim, which was presumably useful in cases where a neighbor built an overhang that extended into the airspace of his neighbor, came to be interpreted by some to mean that state sovereignty extends "all the way."

Another principle, this time derived from traditional international law, is that sovereign territory by definition is that within which a state has the right to make its laws effective to the exclusion of all other states. A literal interpretation of this maxim would mean a different size airspace for every country depending on its anti-aircraft fire power. This theory has had an occasional supporter who would apply to all of space the test of "if you can shoot it down, it has no right to be there." But there remain only a few who will argue that national sovereignty exits as far as it is physically and scientifically possible to control it from below. Oscar Schachter among others has lucidly criticized the essential immorality of the doctrine that "might makes right." In general, it seems a poor basis for a regime of international law, and in fact such rules as do exist presuppose uniform rights in the airspace and undoubtedly in outer space as well.

Also, one hears little more of the "pie-wedge" theory elaborated by Hans Kelsen, to the effect that the zone of national sovereignty extends from the center of the earth out to infinity along lines that cut through national boundaries. This doctrine might have sufficed in a Ptolemaic universe but simply will not fit the facts of astronomy, since the position of the universe comprehended in any country's mythical "cone of sovereignty" changes every instant with the rotation of the earth combined with the natural movements of other celestial bodies. It is more seriously argued that the *usque ad coelum* maxim does carry the "territory" of states up to and including the airspace, and John Cobb Cooper has concluded from this that the airspace is consequently a *part* of state territory (October 10, 1960). Certainly there is little to dispute as to the rights of states in their own airspace. But where does it end, and outer space begin?

THE BOUNDARIES OF SPACE

The atmosphere, as it turns out, does not lend itself to sharp gradations. Indeed, it can be—and has been—argued that the atmosphere extends anywhere up to 20,000 miles above the earth's surface. But half of the entire mass of the atmosphere is less than 3½ miles above the earth's surface; it is below, for instance, the peak of Mt. McKinley. The troposphere, from the surface to approximately 6 to 10 miles up, includes 75 per cent of the atmosphere's weight. An abrupt change in the temperature gradient occurs at the tropopause where the stratosphere begins. The lower stratosphere extends up to approximately 18 miles, which is beyond the area containing 97 per cent of the earth's atmosphere. The mesosphere or upper stratosphere then continues to about 45 miles.

The ionosphere ranges from 40 to 50 up to about 300 miles, but at 50 miles atmospheric density is only one millionth that of the earth's surface. At 100 miles up the temperature ranges up to 2000 degrees F., the density is one billionth of the earth's surface, and oxygen molecules have broken down to atomic oxygen. Finally, the exosphere, beginning at 200 or 300 miles, can be argued as extending anywhere from 800 to 20,000 miles, depending on the importance one attributes to the presence of ionized particles. It is instructive to know that the mean free path—or distance between molecules—which is one millionth of an inch at sea level goes up to 51 inches at 70 miles, a half a mile at 140 miles, and 43 miles at 250 miles.

Another technical complication in drawing neat boundary lines, found by the International Geophysical Year, is that the density of air in the high atmosphere varies by a factor of ten, depending on the geographic position, the time of day, and the season. And a final complication is that the aerodynamic features of flight which result in the flight of airplanes have almost completely disappeared by about 50 to 60 miles—but new hybrid craft will pass back and forth.

So it can be seen that the various proposals made for fixing the boundary lines between earth's atmosphere and outer space cannot rest on unchanging data of scientific certainty. These proposals range from 25 miles to infinity, and some of them appear to make better sense than others.

The 25-mile figure has been favored by some important officials in the United States government as the proper distance to which each nation's sovereign airspace should extend. Quite obviously, a 25-mile boundary would favor experimentation by nations in space with minimum difficulties arising from charges of intruding in the airspace of nations being "over-orbited."

Thirty miles has been favored by some writers, while 52 miles is chosen by others as the "height at which centrifugal forces take over." Andrew Haley has suggested a limit which bears a critical relation to space flight, the "Karman line" (named after Theodore von Karman). This is a curve of altitude plotted against velocity, connecting the points at which aerodynamic flight effectively ends and centrifugal force takes over. Haley defines the "Karman primary jurisdiction line" at the present state of technology at approximately 55 miles. But the X-15 rocket plane has already blurred the distinction; and when the *Dyna-Soar* boost-glide type of vehicle is perfected, it will cruise at speeds of 16,000 to 18,000 miles an hour at altitudes of about 60 miles, carrying passengers half way around the world in 90 minutes.

The lowest height at which artificial unpowered satellites can be put into orbit at least once around the earth is becoming increasingly popular as a possible boundary. This, according to a number of authorities, is somewhere between 70 and 100 miles. But some of the performance figures of satellites show the complications here too. A score of satellites have had a perigee of under 300 miles, and a dozen a perigee of 156 miles or less; *Explorer VII's* was 100 miles. An elliptical orbit brings a satellite very close to earth, but its apogee would be vastly distant, posing the possibility of having a single satellite subject to two completely different legal regimes. For example, *Explorer VI* had a perigee of 156 miles with an apogee of 26,357. The *Pioneer V* space probe of March 1960, with an estimated life span of 100,-000 years, had a perihelion of 75 million miles and an aphelion of 92.3 million miles. The Jodrell Bank Radio Observatory heard information emanating from that craft up to 22.5 million miles from earth. Finally, auxiliary power sources on satellite vehicles will enable them to orbit at ever lower altitudes.

Still other standards have been suggested, such as the height at which the earth's gravitational pull significantly diminishes, the height at which no molecules of gaseous air are found, and so forth. John Cobb Cooper has suggested a variation, involving a contiguous zone between territorial air space and outer space which would extend either 300 or 600 miles between the two. In this zone the right of innocent passage of non-military craft would be maintained in somewhat the same way as in the territorial waters of a state. (The Inter-American Bar Association has even given a name to such a neutral zone: "Neutralia.") A variation of this proposal is that altitude be completely abandoned as a criterion of jurisdiction in favor of the concept of trajectory. The proposal (by Leopold and Scafuri) would exempt from national territorial sovereignty all craft which describe orbital or suborbital trajectories.

And finally there was a surprising agreement at a meeting of the International Aeronautics Federation on October 4, 1960, when United States and Soviet participants agreed to a standard by which space flight records would be judged. This standard was that flights (in this case by manned rockets) would have to reach 62 miles to qualify as space flights. Perhaps that is to be the new magic figure.

There is no consensus yet on the boundary line, but the elements of a consensus are not entirely lacking. It seems to be overwhelmingly agreed by both legal commentators and governments that national sovereignty does cease at some point in space, and that there must be some limit above which national boundaries become irrelevant and space is "free." Those who oppose this position are in a small minority. But it is clear that both science and engineering are continuously complicating the task for the earth-bound lawyer and diplomat, and may continue to do so. Secondly, it seems likely that if a consensus develops it will be on a boundary at somewhere between 25 and 100 miles. Lastly, governments seem less anxious than private publicists to fix such a line.

If it should become possible to negotiate an international agreement defining the boundaries of outer space, several interesting historic points of reference are pertinent. The Paris Convention of 1919 for regulating air navigation stated that "Every Power has complete and exclusive sovereignty over the air space above its territory"; but the air space was not defined. Twenty-five years later, in Chicago, the Convention was brought up to date and signed by all the major powers except the Soviet Union and China. It repeated the principle that "every state has complete and exclusive jurisdiction over the airspace above its territory," and in a later annex it repeated the definition contained in the Paris Convention of an aircraft as "any machine which can derive support in the atmosphere from reactions of the air." It also asserted, with uncommonly low predictive powers, that pilotless aircraft might be flown over states only with their permission. From the background of the air conventions it is argued that national sovereignty extends only into the airspace, and that freedom of outer space is thus established in law. Whether this is in fact so remains to be seen.

SOVEREIGN RIGHTS IN OUTER SPACE

Another issue—national sovereignty in outer space—is already far from theoretical. Two events took place in 1959 that suggested some of the contours of the future, both resulting from Soviet initiatives.

First the Soviet moon shot in September of that year carried in its nose cone numerous pieces inscribed with the hammer and sickle and etched with the characters "CCCP," which were to be scattered on impact, plus a foot-long Russian ensign. But when asked after the successful impact whether the Soviet Union would lay claim to the lunar regions, Alexander D. Topchiyev, Deputy Chairman of the Soviet Academy of Sciences, said "No, . . . there will be no territorial claims." (Moscow, September 13, 1959).

The second event followed within the month. On October 4, 1959, the Soviet Union launched *Lunik III,* which proceeded to photograph the far side of the Moon. *Pravda* on October 27 printed a map which clearly showed that the Soviets were going to claim the right of the ancient discoverers by unilaterally giving names (Russian, naturally) to nearly all the major features of the back side of the moon, thus breaking the time-honored custom of the International Astronomical Union of naming newly-discovered features of celestial bodies.

The United States did not react officially to the Soviet assertion of its right to name that which it had discovered. But after the impacted landing Washington took the position that the placing of national insignia was an insufficient basis on which to found a claim of sovereignty over an unoccupied land mass. In its statement the United States also questioned whether celestial bodies were capable of being appropriated by national sovereignty.

The positions taken by the super-powers regarding their legal rights in space go back to a much earlier body of international law with respect to territorial claims. Perhaps the first attempt to introduce the idea of law into sovereign claims was in the famous Papal Bull of May 4, 1493, in which Pope Alexander VI attempted to divide the New World, including its undiscovered areas, between Spain and Portugal. The papal division was soon succeeded by the so-called "hinterland principle"; he who owned the coast could claim the region inland to an indefinite extent. This formed the basis for the original land grants of the American colonies and the principle for territorial claims in Africa until the last great land rush in the 19th century. In that earlier age of exploration England claimed most of North America because John Cabot in 1497 had sailed along the coast and spent some time ashore. Such claims inevitably led to war, particularly when they rested on such flimsy foundations as the placement of crosses or monuments or the English "turf and twig" symbolic ceremony.

The hinterland principle was in turn succeeded by the doctrine of effective occupation. As articulated by Secretary of State Charles Evans Hughes in 1924, sovereignty requires posession and also occupation with intent to make it permanent. Discovery might give an inchoate title, but it has to be followed up by occupation within a reasonable time or it lapses in favor of any other state succeeding in occupying the area in question. The famous Palmas Island arbitration of 1928 exemplified this doctrine.

In outer space the problem is already before the nations. It will become acute when the first man from earth sets foot on the moon, predictably within the decade. What ground rules should govern the status of the moon and the planets in the light of all this?

Several legal principles have been suggested by contemporary writers. Oscar Schachter, along with others, has proposed that outer space be treated as *res communis,* like the high seas. In other words, it should not be considered susceptible to appropriation by any nation. He suggests that the use of natural resources of the planets be considered analogous to property rights in sedentary fishing grounds, which include the right of innocent passage. Wilfred Jenks defines space beyond the atmosphere as *res extra commercium,* similarly incapable of appropriation by projection of territorial sovereignty. He proposes that sovereignty over unoccupied territories beyond the earth as well as title to natural resources be vested in the United Nations.

The high seas analogy is a logical and useful one, but at the same time is more complex than it appears. The traditional three mile limit, for example, is riddled with exceptions. International law recognizes a qualified right of innocent passage through territorial waters and special rights for fishing and exploitation of minerals on the continental shelf. A four mile limit was recognized by the International Court of Justice in the Norwegian Fisheries case. Since 1790 the United States has claimed jurisdiction for customs purposes in a contiguous zone extending up to 12 nautical miles. The American air defense identification zone goes well beyond these limits. And most recently the United States—unavailingly, as it turned out—offered to compromise on 6 miles at the recent United Nations Conference on the Law of the Sea. Nevertheless, the analogy is a sound one and should doubtless be applied to outer space as a whole.

Another analogy which has been suggested stems from the treaty concerning Antarctica. In that treaty, signed in Washington on December 1, 1959 the states involved, while not formally relinquishing their claims to territorial sovereignty on the Antarctic continent, agreed to place those claims in abeyance, without prejudice to legal rights, and to open the entire continent to exploration, either cooperative or individual, with complete freedom of access for all. Accordingly to the agreement Antarctica is never to be militarized but is to be devoted to scientific study for the benefit of mankind. Antarctica, in other words, has become in fact *res communis.*

Outer space and Antarctica share the happy quality of being, so far as we now know, without human inhabitants. The Antarctic Treaty, with a change of only a few words, could be very nicely applied to outer space, but with one important proviso: control of space requires agreed inspection procedures involving physical presence at launching sites, etc. It would hopefully not be sufficient simply to allow everyone to observe the area freely as in Antarctica.

The United States has taken some leadership in urging the nations to establish basic principles in this realm now, while there is time, before conflicts actually occur on other bodies. Others have held the same view. In May 1958 the UN Secretary General, Dag Hammarskjöld, called on all nations to renounce territorial claims in outer space, basing his reasoning both on the precedent set by the International Geophysical Year and on the analogy to the high seas. In his speech to the UN General Assembly on September 22, 1960 President Eisenhower suggested that the Antarctic principle be extended to outer space, where national vested interests had not yet developed and where barriers to agreement were lower than they ever will be again. He specifically proposed that celestial bodies be agreed upon as not susceptible to national appropriation by any claims of sovereignty. President Kennedy in 1961 proposed that the United Nations Charter be considered to extend to the limits of space exploration and reiterated the American proposed ban on sovereign claims in space.

It is important, before too much further time elapses and too many actual landings are made, as they will be, that a formal international agreement be conducted comparable to the Antarctic Treaty, proclaiming that outer space and the territories therein are considered *res communis,* open to all for peaceful purposes and not subject to national appropriation, and should be reserved for beneficial, knowledge-producing activities. It would be equally desirable to specify in an international agreement that space not be militarized, although of course the problems of inspection would be far more difficult and complex than in Antarctica.

Even before such agreements were actually debated, some principles have been in process of becoming established. Already a kind of common law has developed about "overflights" (although it remains to be seen if the American reconnaissance satellite will cause the Soviets to revise their doctrine). So far, at least, the principle that outer space—wherever it may begin—is essentially "free" and open to all is becoming generally recognized. Ever since October 4, 1957, when *Sputnik I* went into orbit, a precedent has existed for the free use of outer space without objection by the state over which the satellite has passed. But this is only a fortuitous beginning. Western statesmanship has been somewhat inclined to leave these matters to chance, and for a time our position seemed limited to having our legal counsel reserve the rights of their clients to do something unconstructive if someday someone else were to. But—as we discovered when this rather unimaginative view was presented in the United Nations on several occasions in recent years—even our friends found it necessary to tell us that such a position did not do justice to a nation ostensibly leading the way to the preferred version of world order. The United States position in the fall of 1961 went some way toward articulate leadership in this vital realm, but it is only the beginning, not the end, of the process of rule-making for a lawful and rational regime in outer space.

NATIONALITY AND LIABILITY FOR DAMAGES

The remaining legal issues are less far-reaching but possibly of even greater practical importance. An interesting question has to do with the status of space craft. Vessels on the high seas, according to international law, carry so to speak the nationality of their flag. Does a space vehicle carry with it the nationality of its flag, like a ship on the high seas? According to the Paris and Chicago conventions, the principle of nationality applies to aircraft. Thus an accepted rule exists according to which the state of the flag is responsible for the international good conduct of its vessels and aircraft when they are in service beyond the nation's territory. Moreover, a state has in return the right to expect that its own national vessels and aircraft are accorded the rights and privileges by other states to which they are legally entitled.

Normally the "state of the flag" would be the one which had launched the satellite or space vehicle, although an interesting question arises as the United States and possibly others use their own rockets and launching facilities to place into orbit satellites containing payloads owned by other states. But the principle of nationality is an important one and indeed indispensable when it comes to fixing responsibility with regard to radio frequencies, to protesting against specific international regulations, and— perhaps most practical of all—in the matter of liability for damages by or to spacecraft.

The question of damages is considered by some space lawyers to be the most urgent topic for international rule-making. Once again, some principles of law can be carried over from the past to the future in order to develop a doctrine that is both practical and equitable. The principle of absolute liability in Anglo-Saxon law stems from two sources. Under the old common law an individual was liable for any injury which resulted from his trespass or physical entry onto another's land, however accidental. This principle was mainly replaced by the requirement that liability be based on fault. Contemporary doctrine is a mingling of the two: a person who engages in an activity involving undue risk to his neighbors is strictly liable for any injury which may take place even though the activity be strictly lawful.

Another interesting precedent was set in 1941 when an international arbitral tribunal found Canada liable to the United States for damages which resulted from fumes which emanated from a Canadian smelter and drifted across the border. Similarly, in domestic law airlines are frequently made completely liable for any injury they cause to persons on the ground without proof other than that damage was caused by their aircraft in flight or by persons or things falling therefrom. International conventions signed at Rome in 1933 and 1954, though ratified by few states, sought to extend this principle to private international law.[1] It seems highly impractical to wait for customary rules of law to develop, for there is a clear need for international rules determining the rules of liability, perhaps by applying the Rome convention directly to spacecraft.

Proof of fault is required in collisions, of course, and in the case of spacecraft, governments rather than airlines would be liable. An interesting proposal by the prolific Mr. Cooper is that since the mills of international claims, lawsuits, and diplomatic negotiations grind exceeding slow, it would be helpful if each state created a guarantee fund to compensate its own nationals and then accepted the compulsory jurisdiction of the International Court of Justice so that the state could eventually recover from another state. The same principle was suggested by the Council of the Inter-American Bar Association in its so-called "Magna Carta of Space" voted at Bogotá on February 3, 1961, which included *inter alia* a recommendation for an international insurance fund to provide for compensation. This seems a most reasonable suggestion, particularly where one contemplates endless litigation spanning decades and resulting not only in inconvenience but also in exacerbation of international relations. There is of course the path taken by the United States in its *ex gratia* payment of two million dollars to the Japanese fishermen who were unaccountably injured in the hydrogen bomb tests in the Pacific. It would seem far more sensible to create an international mutual insurance fund to take care of losses resulting from accidents in and from outer space.

METALAW

One final aspect of international law on celestial bodies is the very human—if one can use that word—issue of the rights of other sentient beings on celestial bodies which *homo sapiens* might stumble upon. Andrew Haley has suggested a law which spans both our civilization and theirs, and which he calls Metalaw. It is based on the motto, "Do unto others as they would have you do unto them." Certainly it would be consistent with contemporary principles to acknowledge that earth people have no right of owner-

---

[1] For all this information I am grateful to Eric Weinman and Hugh C. MacDougall; see *Foreign Service Journal,* April 1958.

ship or control of any other inhabited planet; the far more tenable assumption is that such a planet is owned by the beings—if any—who live there.

*   *   *

There is clearly no general agreement, even among the Western students of the problem, on what is important and what is unimportant in the realm of space law or on what the content ought to be of the laws that are important. This, in turn, poses the far more sobering problem of establishing a consensus between Western and Communist powers on principles of law with regard to space. Some matters might well be negotiable, both among ourselves and with our global adversaries. These might include such intensely practical matters of mutual interest as the creation of a mutual insurance fund for compensating damages, the allocation of radio frequencies, provisions guarding against contamination and requiring that dead radios be turned off, return of parts of satellites, and so forth. It is in the realm of more abstract issues—the definition of boundary lines, the principles of territorial sovereignty, and above all the demilitarization of outer space—that the clashes of ideology, of prestige, and of national power impinge heavily on the process of rule-making, just as they do with respect to sensitive issues on the earth.

In such negotiations it is not necessary for governments to forego any rights which they might later want to claim. Our government should however build on its present initiative to propose and secure a broad-spectrum international agreement spelling out ground rules for space which, if generally adopted, would require all nations to forego claims that make little sense anyway. The American method of operation tends to prefer the step-by-step, pragmatic approach to the more grandiose approaches to rule making. This was not, however, true with respect to the creation either of the American Constitution or the Charter of the United Nations. The United States can continue to decide upon and clearly enunciate, in appropriate detail, the principles which could begin to set the tone for acceptable international behavior, principles which not only deal with functional areas, such as allocation of radio frequencies, liability for damage, protection against contamination, and the like, but also include definition of the outer boundary of national airspace, with a neutral contiguous zone between that and the start of outer space proper; the proposition that space, the moon, the planets and other heavenly bodies are *res communis* not *terra nullius,* and, by analogy to the high seas, open to peaceful use by all without regard to nationality and not subject to national appropriation; the statement that we are ready ourselves to negotiate and sign an outer space treaty in part analogous to the Antarctica agreement; and further development of the international functions of registering the launching, orbital, and other identification and performance data, as well as that of clearing house for the technical and scientific information made available.

The Political Problem of International Controls

PREFACE

Three facts stand out in considering how the space age might be brought within some kind of international framework to ensure that it becomes a force for human good and not simply another source of human anxiety and, ultimately, destructiveness.

First, the space age is only a few years old. In the annals of history this is a miniscule segment of time in the process of building communities and, particularly, institutions on a world-wide basis.

The second fact that stands out is that space technology is extraordinarily dynamic and proliferating at such a rate that unless some international rules are agreed upon, the consequences will be, at best, chaotic, at worse calamitous. Third, even the tiny steps which the world community has already taken in this field are intelligible only against a backdrop made up of the political problems which man on earth has generated. These are the great and seminal problems of nationalism of ideology, of competing theories of society, and of historic failures to create supranational institutions on other grounds; man carries all of them with him into space.

Against these three sets of facts—the short history, the fast moving technology, the earth-bound ideological wars—the issues of international control take their real meaning. The efforts so far to find handholds on the problem have taken two basic forms. One grows out of the category of peaceful uses of outer space—the utilization of space technology for communication, for weather research and forecasting, for navigational aids, for scientific investigation of the earth's shape, of the atmosphere, and of space itself, and, finally, for the exploration of the moon, the planets, and who knows what extra-galactial phenomena. The other great category has to do with the military use of outer space. These two categories—peaceful and military—define the kind of approaches which can be made to the problem.

One approach focuses on the prohibition of outer space for war-like uses. It is this approach, for example, to which disarmament and arms control proposals have tended to address themselves. Starting with 1957 and continuing through the various UN General Assembly sessions as well as the disarmament negotiations, proposals have been advanced by both sides calling for the banning of outer space to military uses and ensuring that it be used for peaceful purposes only. The United States has since then become more specific in urging that a ban be placed on the stationing of nuclear devices in orbit. In general the prohibitory approach has been limited to the disarmament area. Dr. Brennan's chapter discusses the substantive issues involved in these efforts.

The other approach—the more positive and affirmative concentration on peaceful uses and the encouragement and regulation thereof—has been the principal focus of United Nations activities.

THE UNITED NATIONS SPACE COMMITTEES

In December 1958 the United Nations General Assembly on motion of the United States created the Ad Hoc Committee on the Peaceful Uses of Outer Space, consisting of 18 members.

The negotiations for the creation of the Ad Hoc Committee illustrate vividly the interconnections between the space problem and the cold war. At first the Soviet Union made as a condition the elimination of foreign bases as noted above. As the negotiations for the resolution continued, however, the Soviets dropped their insistence that the issue of foreign bases be tied to the space item. They also dropped their proposal for a United Nations Agency; and the Western powers, for a reason which is quite inexplicable to the writer, in advancing their own proposals neglected to pick up this theme themselves.

The most egregious difficulty, however, arose over the problem of "parity." Just as the Soviets later demanded parity in the office of the Secretary General, so even in 1958—a year after *Sputnik I*—they were already insisting on greater recognition in the composition of the UN's first outer space body for their presumably preeminent position in the field. The composition was not negotiated to Soviet satisfaction, and the Soviet Union, followed by Poland, Czechoslovakia, India, and the United Arab Republic refused to take part in the work of the Committee. The Committee went ahead nevertheless, dividing itself into two committees of the whole, one legal and one technical; and it held 25 meetings in all.

It is generally considered to have been a fairly rudimentary exercise, and one which was extremely inhibited, both by the absence of the Soviet Union, and, it must be said, by the politically incomprehensible conservatism of the United States on this subject. The final report was adopted on June 25, 1959 by the thirteen remaining states. It was clear that the inhibiting effect of the absent Russians made the Committee more cautious than even the protective and conservative tone of American policy at that time justified. Yet despite the handicaps the Ad Hoc Committee represented the first—indeed the only—intergovernmental attack to date on the problems of outer space at the political level.

The Committee had been asked by the General Assembly to report on, among other things, "the nature of legal problems which may arise in the carrying out of programmes to explore outer space." While the Committee considered that some legal problems were "more urgent and more nearly ripe for positive international agreement" than others, it concluded that a comprehensive code was "not practicable or desirable at the present stage of knowledge and development." Its report pointed out that relatively little is known of the uses of outer space, that the rule of law is not dependent on comprehensive codification, and that "premature codification might prejudice subsequent efforts to develop the law based on a more complete understanding of the practical problems involved."

(This position is remarkably similar to that taken by the American Bar Foundation, which on December 5, 1960 announced that it would be "premature and perhaps even harmful" to attempt now to draft a comprehensive space law because so little is known about the uses of space. The effort "might come to naught or yield a small set of pious maxims of extreme generality, or produce an unworkable regime that would be all the more dangerous for giving the temporary illusion of certainty.")

The Ad Hoc Committee did agree on the need to take some timely and constructive action. It proceeded to group the basic problem areas according to priorities, and it suggested that the Assembly keep these problems under review in the future.

In the eyes of the Ad Hoc Committee the following legal problems were deemed susceptible to priority treatment. First, the question of the freedom of outer space for exploration and use. Here the Committee enunciated the first authentic statement of the principle of freedom of outer space. On the basis of the practice of the International Geophysical Year the Committee believed that, regardless of what territory was passed "over" by space vehicles, there may have been initiated the recognition or establishment of a generally accepted rule that "in principle, outer space is, on conditions of equality, freely available for exploration and use by all in accordance with existing or future international law or agreement."

Next in priority was the question of liability for injury or damage done by space vehicles. Without answering any of the questions it raised, the Committee recommended early consideration of an agreement on submission to the compulsory jurisdiction of the International Court of disputes between states regarding liability.

The Committee also assigned priority to the question of allocation of radio frequencies, the avoidance of interference between space vehicles and aircraft, the identification and registration of space vehicles and coordination of launching, and, finally, the problems surrounding reentry and landing. The Committee cited the need for multilateral agreements providing for cooperation in returning reentered vehicles to the launching state; it suggested that existing rules regarding aircraft landing on foreign territory in case of accident, mistake, or distress be applied here.

The Ad Hoc Committee assigned to the category of "non-priority problems" the question of where outer space begins. With "no consensus" they could discern, the members of the Committee concluded that an international agreement now would be "premature." They suggested the possibility of tentatively setting a range with a boundary not so low as to interfere with existing aviation nor so high as to fetter exploration activity.

Other non-priority problems involved public health and safety, the exploration of celestial bodies, and the avoidance of interference among space vehicles. It can only be said that neither the order of priorities assigned by the Committee nor its excessively cautious approach necessarily represent a wide consensus among other professions concerned with the creation of order in outer space. Nevertheless, a start had been made.

When it turned to the broader question of institutions and controls, the Ad Hoc Committee threw little light on the problem. It considered it "inappropriate at the present time to establish any autonomous inter-governmental organization for international cooperation in the field of outer space" or "to ask any existing autonomous inter-governmental organization to undertake overall responsibility." But its Technical Committee sensed a need for "a suitable centre related to the United Nations that can act as a focal point for international cooperation in the peaceful uses of outer space." The Committee suggested that the Secretary General organize a "small expert unit within the Secretariat" for this purpose.

After the Ad Hoc Committee had made its report in June, 1959, the outer space issue was debated once more in the General Assembly in the fall of that year, and a resolution was this time unanimously passed. Entitled "International Cooperation in the Peaceful Uses of Outer Space," this resolution made no reference whatever to the work of the Ad Hoc Committee, presumably because the non-participants did not recognize the latter's legality. In the new resolution the Assembly recognized the common interest of mankind as a whole in furthering the peaceful use of outer space and registered the belief "that the exploration and use of outer space should be only for the betterment of mankind and to the benefit of states, irrespective of the stage of their economic or scientific development." It established a brand new Committee on the Peaceful Uses of Outer Space, a committee which accepted a principle vulgarly known as "soft parity" in which 12 Western or pro-Western states were balanced against 7 Communist bloc states and 5 neutrals.

The resolution requested the Committee to "review . . . the area of international cooperation, and study practical and feasible means for giving effect to programs on the peaceful uses of outer space which could appropriately be undertaken under United Nations auspices." These included assistance in the field of research, information, and the study of legal problems. The resolution called for the convening in 1960 or 1961 of an International Scientific Conference under the auspices of the United Nations for the exchange of experience in peaceful uses.

By late 1961 the new Committee had still not met. Instead, it has been stymied on a set of new issues regarding officers of the committee and of the scientific conference and the voting procedure in the Committee itself. In the establishment of a committee on arrangements for the scientific conference the Soviet Union proposed a composition of three Western representatives and three members of the Soviet Bloc, with a Soviet Chairman. The Russians also proposed a Soviet Secretary General for the scientific Conference. Mr. Zorin was quoted on April 25, 1961 as testifying that Soviet leadership in space exploration entitled it to persist in its demand for "a footing of equality" in the Committee. The Soviets reminded the United States that they had agreed to an American as Secretary General of the 1955 United Nations Conference on the Peaceful Uses of Atomic Energy and also to a United States national for director general of the International Atomic Energy Agency.

However, the United States rejected this Soviet proposal. The Soviets then came up with a proposed committee on conference arrangements consisting of two Western states, two Soviet bloc states, and four neutrals, with the Soviets still as chairman. The United States did not accept this arrangement, reiterating its conviction that a neutral should be chairman. The Soviets agreed to the United States proposal that the scientific and legal committees of the new Committee should be committees of the whole, but no agreement had been reached on the other issues by the fall of 1961. For the chairman of the new Committee the United States backed Madame Rossel of Sweden, while the Soviets sponsored Mr. Jha of India.

It was clear that the Soviet Union, while accepting soft parity in the composition of the new committee, wanted what it considered to be real parity, and indeed recognition of what it considered its actual leadership in space technology. It was understandable but not entirely commendable that the United States government, having accepted leadership in the atomic energy field when it had pre-eminence, should adopt the position that there should be *no* great powers in positions of authority either in the committees or in the scientific conference. We do not wish to exaggerate the Soviet lead in space, nor do we want to seem to abandon our preferred power position in international organization. It would, however, seem reasonable and even graceful to suggest that a Soviet scientist chair the scientific conference on space. Another procedural issue that blocked the committee from functioning had to do with voting arrangements. Under General Assembly Rule 162, a simple majority generally prevails in an Assembly Committee. In the new Committee the Soviets sought to establish the principle either of unanimity or of the two-thirds vote. This was a poor idea, and the United States would do well to continue to oppose it.

It is interesting to note that the same struggle has gone on in the nongovernmental organizations—the International Astronautical Federation (IAF) and the COSPAR—which is referred to in Dr. Odishaw's paper.

Since all of them are places where the Soviet leaders believe they have a good claim to parity and even to positions of leadership, it seems unlikely that a genuine compromise can be reached unless the United States acknowledges the Soviet claims, or, preferably, somehow restores the balance of technological power to the point where the Soviets may claim parity but not leadership.

Regulation of Functional Activities

Quite apart from the issue of political controls and institutional regimes to encourage and sponsor peaceful uses or to enforce a prohibition of nonpeaceful uses, there is a wide range of functional activities suitable for international regulation or, conceivably, even international operation. These include allocation of radio frequencies, weather reporting, scientific dataprocessing and dissemination, and the like. As in the UN system generally, functional cooperation is not only possible but is actually taking place within and outside the UN family of agencies despite the unedifying impasses and power struggles in political organs which continue to mirror the larger political battles outside the United Nations.

In its report the Ad Hoc Committee reported the findings of its technical committee. The latter spoke of the need for international arrangements for the orderly developments of scientific and technical problems of space. An example is tracking stations, which have to be located in many places on the earth. As the Technical Committee pointed out, "no single country extends over a sufficient range of latitude and longitude to be able to track earth satellites adequately from its own stations." The Technical Committee said that "radio interference from terrestrial sources could cripple the conduct of space programs." It pointed to other areas where cooperative action was needed.

Some of these problems are being attacked by inter-governmental organizations, while others remain to be tackled. The sections that follow include references to the work of the UN specialized agencies, some of which are already at work on the undramatic functional problems raised by the space age, without regard to the great strategic and political maneuverings which so far preclude agreements on a more sweeping basis.

REGULATION OF RADIO FREQUENCIES

In this field some progress has indeed been made. The Ad Hoc Committee's report, in giving priority to the matter of allocating radio frequencies, recommended that a link be established with the International Telecommunication Union, one of the UN Specialized Agencies. The International Radio Conference of the ITU in 1959 acted on American initiative to allocate radio frequency bands for remote control space vehicles, and specific allocations for space and earth science space radio services were included in the radio regulations annexed to the International Telecommunications Convention signed in Geneva in 1959. This was the first concrete regulatory measure in the outer space field. It is interesting to note that 43 space vehicles containing over 53 radio transmitters were launched in orbit or on space probe missions before there was a binding rule on the use of the radio spectrum in astronautics.

The ITU now has under review future needs in this field, through its International Radio Consultative Committee (CCIR) whose job it is to study technical questions regarding radio operations and make recommendations. It meets every three years, and its work is carried on in the interim by study groups. Problems are multiplying in this field as the numbers of space vehicles increase and radio equipment is put to new uses. Future needs involve the identification of radio emissions from space vehicles, codes for the transmission of information to space vehicles, methods of providing for the cessation of radio emissions from space vehicles, and space vehicle relays to extend terrestrial communication.

All nations involved have tentatively agreed to convene an Extraordinary Administrative Radio Conference to be held, possibly in 1963, to provide adequate radio spectrum segments for all categories of space communications. It will be convened if sufficient data regarding the means and manner of use of additional frequencies are delivered to the administrative council of the ITU by May 1962.

With the United States' proposals in the fall of 1961 calling for planning and action within the UN's system in the field of satellite communication, the prospects had increased for eventual development of a well-integrated global system of communication satellites linking the world by telephone, telegraph, radio and television.

All in all, this has been so far a heartening episode of international cooperation. But new problems are going to arise. What, for example, will other countries get out of agreements on the international allocation of frequencies which the United States needs? Here one must face up to the lack of symmetry in the development of space technology; for the Soviet Union and the United States require that bands be set aside on the radio spectrum, in some cases at the expense of other countries. Of course, global communications may be established which may offer something tangible to other countries which cooperate. And there would also be a political quid pro quo in the establishment and operation of an international agency in which other countries have an appropriate voice.

INTERNATIONAL WEATHER SERVICES

The World Meteorological Organization—another specialized organization of the United Nations—is playing an active part in advising on artificial satellites for meteorological purposes. It has had discussions on the role it can play in planning tracking systems and in the design of computers to process meteorological data. In 1959 it set up a panel of experts which suggested that coordinated facilities were needed for interrogating weather satellites, rapidly reducing data received from satellites in a form that would be useful for synoptic meteorology, and for a systematic world-wide exchange of data for immediate use. It has also discussed the possibility of entrusting the WMO itself with operational responsibilities. In May 1961 the WMO executive committee met in Geneva and heard United States representatives describe the results of the *Tiros* shots. Other governments expressed great interest in having the advantage of access to the information received from our system. The Soviet delegation was reported to be cooperative and interested in discussing international cooperation in this field—as well it might be, considering the land mass covered by the Soviets and the need for reliable weather information for a variety of purposes. But only time will tell whether this interest, or the Soviet fear of Western espionage through *Tiros* observation, will predominate in Soviet policy.

The United States Weather Bureau distributed pictures from *Tiros* through the normal channels to the WMO, and it was the United States that called the meeting of some 100 other nations including the Soviet Union to cooperate in a new weather-forecasting system based on the prospective *Nimbus* meteorological satellite. Here, as in the radio field, vistas have opened up for international cooperation both on a global level and in terms of regional research centers. As the *Tiros* system becomes operational it is a matter of both necessity and common sense that the information, and perhaps, ultimately, the systems themselves, be internationalized; weather information is a reciprocal process the world over, and the greatest common interest is served by making its results available on a world-wide basis. United States proposals in late 1961 could hopefully yield coordinated international programs and action within the next few years.

OTHER INTER-GOVERNMENTAL ACTIVITIES

Other activities are going on in the UN family of agencies. UNESCO—(the United Nations Economic, Scientific and Cultural Organization)—has an agreement with the International Council of Scientific Unions (ICSU) involving the stimulation of scientific research in the space field, both globally and through regional arrangements. UNESCO has agreed to undertake only those requests for space research which COSPAR—the space research arm of ICSU—cannot itself do, and these, of course, require intergovernmental agreements.

Another example of cooperation between inter-governmental and non-governmental international organizations was the sending of an observer by COSPAR to the 1959 ITU conference in Geneva. UNESCO has recently adopted a program which authorized its director general to study nine scientific fields including the "exploration of extraterrestrial space."

Other UN agencies have a special concern for some aspect of the space field. The International Civil Aviation Organization is concerned with seeing to it that there is no interference between aircraft activities and space activities; the World Health Organization and the International Maritime Consultative Organization are involved in problems of health and contaminations, and on developing systems for navigation by satellites.

The Diplomatic Task

So long as space technology remains in the hands only of the two superpowers—the United States and the Soviet Union—there is little chance that the problems of outer space will become disentangled from all the consequences of the political relationships between the two powers. To be sure, there are other countries which are developing space technologies. Britain, France, and Israel have all developed rockets, and it is estimated that a dozen or more countries will have rocket capabilities in the next few years. In the most important sense space is still a bilateral problem, but it is also one which effects the rest of the world and every man, woman, and child in it. I suggested earlier some features of a "common law" which might be said to have developed even as between the two protagonists of the space age. Although neither the United States nor the Soviet Union has yet charged that its sovereignty has been infringed by passage overhead of satellites placed in orbit by the other side, grounds for a conflict of interest are rapidly developing, particularly with the development of the reconnaissance satellites by the United States, and charges of "aggression" are beginning to be heard. If and when nuclear weapons are placed in space the conflict could become acute. It is useful to review briefly the positions both sides have taken on the problems of space.

SOVIET POLICIES ON SPACE

The Soviet position has undergone some tactical changes, but in the main it has consistently expressed the policy of a society dedicated to perpetuating its own power and intensely suspicious of the rest of the world. In December, 1926, according to Robert D. Crane, V. A. Zarzar (who became the head of the Soviet civil aircraft effort) opposed control of space by international organizations dominated by anti-Communist states, he asserted that full sovereign rights should extend through the air space, and limited rights beyond it and as far into outer space as national security interests might require. In 1934 Ye. Korovin added that any attempts to restrict sovereignty in space would be utopian without disarmament and elimination of war. "No responsible Communist writer," says Mr. Crane, "has deviated significantly from these principles in the 30 years since."

But there have been variations on the theme. In the early 1950's the theory of *usque ad coelum* was popular. On the grounds provided by this doctrine of absolute sovereignty up to the heavens, the Soviets protested the 1956 American weather research balloons which were drifting across Soviet borders courtesy of the prevailing winds at 80,000 feet. The then Secretary of State John Foster Dulles' reply implied that legitimate and safe scientific research should not be prevented by nominal claims of sovereignty!

With the launching of *Sputnik I* in October 1957 the doctrine of *usque ad coelum* had obviously reached the end of its usefulness for Soviet policy. Instead the *mare liberum* analogy was invoked for altitudes above about 12 to 18 miles. By January 1959 the theory of freedom of outer space for all countries had apparently been officially adopted. In September 1958 A. Galina wrote that no government should be allowed to incorporate any portion of interplanetary space under its jurisdiction, but argued that rockets could be fired into space without asking the permission of any other government. This very permissive view carried through the launching of the Soviet moon shot in 1959.

Soviet research on legal problems in space, as with other Soviet legal studies, has been completely politically oriented; indeed, Soviet writers are explicit that legal matters in this area have become political. But the apparent American lead in development of observation and reconnaissance satellites is bringing into sharp relief the several strands in Soviet thinking. *Samos* will be a means of outright observation in detail, and *Midas* is designed to improve the Western deterrence posture by detecting rocket exhaust flames sufficiently early to increase warning time by 15 minutes. As a means of equalizing the past American disadvantage in terms of information available about the other side, the American reconnaissance systems are doubtless seen by Soviet military planners as a grave threat. Revelation of relative weaknesses in Soviet offensive and defensive capabilities could have far-reaching consequences for the success of Soviet strategies.

Consequently it was not surprising when in 1959 Mr. Korovin wrote that the development of a reconnaissance satellite would be an expression of mistrust by the United States, although it would not be regarded as an act of war. In July of 1960 two Soviet writers attacked the United States for using the *Discoverer* satellite program to create evidence of "customary norms of law" that would sanction the use of "provocative" and "aggressive" reconnaissance satellites. In the fall of 1960 G. P. Zhukov writing in *International Affairs* took the position that all espionage satellites are illegal and would be regarded *prima facie* as having aggressive purpose:

"From the viewpoint of the security of a state, it makes absolutely no difference from what altitude espionage over its territory is conducted . . . Any attempt to use satellites for espionage is as unlawful as attempts to use aircraft for similar purposes." He went on to say that "in case of need the Soviet Union will be able to protect its security from any encroachments from outer space just as successfully as it is done with respect to air space."

In November 1960 G. P. Zadorozhni writing in the *Stuttgarter Zeitung* said that national sovereignty did not extend to an area where "an artificial satellite can fly freely and not be forced to earth by air resistance" but went on to say that "every nation must have the right to prevent espionage in outer space." And in late July 1961, within a fortnight of the launching by the United States of *Tiros III* and *Midas III,* the Soviet armed forces newspaper *Krasnaya Zvezda* denounced both as acts of espionage and aggression, saying "A spy is a spy no matter at what height it flies." The Soviet position in general is that satellite observation is objectionable not as a violation of sovereignty but as an act of aggression. It is true that it does perform a military function. The *Tiros* program, however, could be—and has been—interpreted by Soviet scientists as one in which the Soviet Union could profitably cooperate. Perhaps they still will.

But in general the ground work has consistently been laid for whatever responses the Soviet Union may make to a successful *Samos* system, which *is* explicitly designed to observe and report back in photographic detail on ground activity. The extent of Soviet sensitivity was revealed in the U-2 episode of May 1960. The fact is, however, that American satellites at an altitude of 200 to 300 miles will eventually do some of the things of which the U-2 was capable. Whether the Soviet objection to a so-called "Spy in the Sky" could lead to retaliatory action cannot be predicted with any confidence. If the Soviet Union were capable of intercepting and destroying satellites in orbit—a capability for which there is no evidence yet—it is possible that this action would be undertaken in the name of Soviet security on the legal grounds which Soviet writers have already advanced. The American reaction, in turn, is also unpredictable. It is conceivable that by the time this became an issue either countermeasures would have been developed, or the United States would be technically capable of orbiting a sufficient number of satellites so that observations could be continued despite such countermeasures.

UNITED STATES POLICIES

Although the United States has played an active role in inviting international consideration of the problems of outer space, as a nation we are not yet entirely clear as to what course in the international arena our national interests truly dictate. Essentially our government seems to have been divided between two courses, both arguably in the national interest.

One views outer space as simply another area in which military superiority is essential; the other sees it as above all a challenge to our capacity to lead the international community toward more orderly procedures.

The first view is not necessarily a militaristic one. Except for the always present lunatic fringe, responsible military leaders regard the military applications of space as basically defensive, as with most of our military preparations. Above all, the *Samos* system of reconnaissance satellites is seen as offsetting in a vitally important way the existing imbalance between American and Soviet information with respect to each other's defenses. In this interpretation *Samos* becomes an equalizing and therefore a stabilizing influence (although the Soviet Union could hardly be expected to so view it), and the function of legal theory here is to protect our right to legitimate defense activities in space and to cover the United States against unforeseen contingencies.

On the other side of the scale, however, is the belief held in important official quarters that the most important requirement is political. The conviction here, widely shared outside the government, favors United States leadership in proposing to the international community a regulatory regime to which the United States would subject itself, conditional upon others doing the same. This group would doubtless advocate far more comprehensive attempts in and out of the United Nations to make rules, create institutions, and to the greatest practical extent internationalize the problem of outer space.

The history of the American position tends to illustrate the interplay of both influences in American policy making. In January of 1957 Ambassador Henry Cabot Lodge presented the first committee of the UN General Assembly with a memorandum on disarmament which included a proposal that future outer space experiments be placed under international supervision and restricted to peaceful purposes. The first step he proposed was to bring the testing of satellites and long-range missiles under international inspection and participation. On August 12, 1957 the West submitted to the disarmament subcommittee in London a working paper including a proposal establishing a technical committee to study the design of an inspection system to ensure that the sending of objects into outer space would be exclusively for peaceful and scientific purposes. The Soviets turned down the proposal, but it was repeated by the United States to the 1957 General Assembly. President Eisenhower several times declared the United States willingness to enter any reliable agreement which would mutually control outer space missile and satellite development. In 1958, in correspondence with then Soviet Premier Bulganin, Eisenhower offered to discuss the problem, but without effect.

A more negative attitude seemed to prevail when it came to action in international bodies. The State Department Legal Adviser Loftus Becker, in Congressional testimony, took the position that the United States reserved its right to make claims in outer space if the need arose and said: "I do not think we should be in a hurry to delimit or restrict the sovereignty of the United States unless we know that we are on the right road." The Congress itself seemed to be more alive than the administration to the desirability of a less negative position. By a unanimous vote the Congress in the summer of 1958 passed a resolution declaring that the United States should seek through the United Nations or other appropriate means an international agreement banning the use of outer space for military purposes and providing for joint exploration of outer space, study of the method by which disputes which arise in the future in relation to outer space be resolved by legal, peaceful methods rather than resort to violence, and an international agreement providing for joint cooperation in the advancement of scientific developments in this field. In May of 1959 the House Committee on Science and Astronautics recommended that the United States should formulate "a positive national policy on the control and use of outer space by the world community." After soberly balancing the pros and cons the Committee concluded that "a more definite policy toward the control and use of space is warranted and that the United States should assume leadership in this phase of the space endeavor."

In his first State of the Union message on January 30, 1961, President Kennedy invited all nations including the Soviet Union to join with us in a variety of space activities, saying that "both nations would help themselves by removing their endeavors from the bitter and wasteful competition of the cold war." And in his address to the UN General Assembly on September 25, 1961, Mr. Kennedy proposed, as part of the U.S. disarmament plan, "keeping nuclear weapons from seeding new battlegrounds in outer space." He further outlined some United States proposals regarding peaceful uses of outer space which included those of "extending the United Nations Charter to the limits of man's exploration in the universe, reserving outer space for peaceful use, prohibiting weapons of mass destruction in space or on celestial bodies, and opening the mysteries and benefits of space to every nation." (I have cited elsewhere the detailed U.S. proposals regarding international weather prediction and communication satellites programs.)

The problem of the reconnaissance satellite remains a sensitive one, however, when it comes to actual international agreements. The United States in its disarmament position has called for the restriction of outer space for peaceful purposes only, specifically urging the banning of weapons of mass destruction, but the reconnaissance satellite in the American view exists for peaceful and defensive purposes, and short of a comprehensive arms control agreement this country is unlikely to agree to prohibiting its use.

THE PROSPECTS

Clearly the space issue between the United States and the Soviet Union is a function of the other political and strategic issues that divide the two centers of world power. It is intimately involved in both the problem of disarmament, including the issue of control and inspection, and the whole issue of building world institutions in the face of Soviet assertions of unqualified national sovereignty. In both areas efforts to bring space under a workable regime collide with the Communists' intense suspicion of any international body endowed with power, which purports to be neutral between the two sides—a concept which we quite correctly favor and they increasingly reject.

While on the part of governments there have been only modest signs of common interest in using space for peaceful purposes and ensuring its freedom for all nations, the private sector has shown considerable initiative in formulating concrete recommendations for the establishment of rules which would bring the space activities of all countries within an international framework. The International Astronautical Federation has pioneered in this field, in 1960 establishing its International Institute of Space Law. The American Bar Association has been similarly active, as have the Inter-American Bar Association, the International Law Association, the American Society for International Law, the Institute of International Law, and numerous other bodies.

The questions which these studies and discussions pose for the thoughtful citizen are basic ones. One of the most basic, in considering both the problems of space law and the question of international political and institutional controls, centers on the pace and the comprehensiveness of official action: how far, how fast, how soon?

The legal world for its part is by and large divided into two broad schools of thought as to how to proceed to attack all the unresolved legal questions. One school of thought advocates a pragmatic, case-by-case approach, in the belief that there is insufficient scientific knowledge on which to freeze such decisions as where outer space really begins. Supporters of this general view feel that premature rule-making would reduce rather than enhance national security: they envisage the gradual development of common law out of practice, experience, and developing consent. This school probably still predominates within the American military establishment, if not any longer in the Department of State. It has considerable support from publicists and scholars, though perhaps not so much as the other school.

The other school believes that the lack of agreed rules regarding outer space can lead to grave international misunderstanding if allowed to persist. In this view legal uncertainties are believed to endanger national and international security in the face of rapid technological change. Apart from the arms race, unhappy consequences of a *laissez faire* policy in space are, for example, the implications of wrongly identifying objects in space, the increasing littering of space, the lack of agreement on ending radio transmissions on dead satellites, the possibility of contamination of the moon or other planets, fogging of photographic plates, and so on. The problem is already with us: the Vanguard satellite launched March 17, 1958 is expected to last 200 years in space with its solar batteries. The conclusion drawn by this second school is that the rule of law must be applied to outer space; its area must be determined, its legal status agreed upon, the rights of states in outer space universally acknowledged, the legal status fixed of flight instrumentalities to be used in outer space. John Cobb Cooper asserts that it would be "disastrous" if no international agreement is reached on these matters (October 10, 1960).

In the realm of political controls, is it realistic to conceive of an international agreement on the status of space to which the conflicting powers in the world would subscribe themselves? Given the interconnections with the disarmament question, can cooperation be continued without having a negative effect on security? Or might it improve the prospects for political agreement? These are the kinds of question that are being asked and to which the answers are slow in emerging.

Apart from the problems of military security—which are profound ones —even those favorably disposed to the notion of de-nationalizing the space race to the maximum extent feasible are divided between those ready to act now and those who consider action premature. The former hold that treaty drafting and institution-building should not wait either for the cold war to end—which may be a very long time indeed—or for all the returns to come in from science and engineering—which will be an equally long time. This school of thought, arguing from the imperative nature of the problem and the growing lag between technology and political arrangements, concludes that a desperate need exists to attempt to formulate some rules and reach agreement before the problem gets hopelessly beyond us, and even before the ideal organizational plans reveal themselves to us.

It is not necessary to think of international ownership of satellite technology in the style of the Acheson-Lilienthal-Baruch-UN Plan which contemplated international ownership of atomic energy facilities. Many believe that the technical problems alone supply a powerful foundation for cooperation in terms initially of a clearing-house function for the appropriate intergovernmental agencies to which would be assigned such areas as communications satellites, frequency allocations, weather information, navigation aids, advance notification of launchings, identification and registration of satellites in orbit, exchange of technical experience and scientific information, agreements regarding re-entry, emergency procedures, etc. (The provisions on identification, registration and advance notice would of course provide useful experience towards any eventual arms control agreement.)

Even among those favorable to stepped-up multilateral activity, some believe that a single UN agency would not be a good idea. The experience of the International Atomic Energy Commission, in this view, argues against such a proposal. Instead it is believed useful to concentrate on individual problems such as communications, beefing up the International Telecommunications Union, or strengthening the World Meteorological Organization to handle increased duties arising out of development of weather satellites, as proposed in the fall of 1961 by the United States. The communications field is a natural for multilateral activity; the widespread benefits to be expected in the near future from a truly international communications system—however sponsored—point clearly to cooperative international programming and regulation. One can conceive either of a greatly refurbished ITA or of a new and separate agency to which the most interested countries would be parties on an open-ended basis and which would be progressively operated by all. This internationally regulated system could be a prototype for a larger sphere of international cooperation.

Beyond both these general positions lies a more extreme kind of proposal. Its proponents consider that measures short of international operation are inadequate, and that even with effective international controls, all rocket ships in the air space would have a military potential and be convertible in time. Messrs. Jessup and Taubenfeld in their excellent study discussed the possibility of international ownership of satellites; exploration and development to be carried out through a Cosmic Development Corporation (CODEC); UN Trusteeship over celestial bodies reached by man; concession and licensing to states of rights in space; and other issues involved in direct international administration.

My own view of what the United States might best do to advance both the national interest and the common interest of all remains as I recently expressed it: "The most effective way to convert the image of American weakness to one of superiority would be by transferring the contest to grounds of our own choosing and daring the Soviet Union to compete with the United States in the degree of imaginativeness and creative statesmanship rather than exclusively in the degree of thrust." [1] The thought was that unless the United States defines sharply, in concrete institutional terms, the image which it wishes to imprint on the space age, the direction which collaboration takes is going to be fixed by Soviet rather than American planners. I believe that the United States should go beyond its present position in proposing the maximum amount of international coordination of space efforts—whether the Soviet Union agreed at this stage or not. The functions of the UN would go beyond those now proposed for a technical office in the Secretariat to include the advance registration and identification

---

[1] Lincoln P. Bloomfield, *The United Nations and U.S. Foreign Policy* (Little, Brown, 1960), p. 227.

of launchings as well as international tracking, monitoring, and reporting of space activity. Finally there would be UN rather than national sponsorship of probes involving other planets and celestial bodies.

These proposals would not tie the hands of our military strategists any more than the 1953 Atoms for Peace Program, which the Soviet Union ultimately joined after the United States initiative won general favor. If 50 or 40 or even 30 nations decided to harmonize and elevate to a multilateral plane their efforts bearing on the peaceful uses of outer space, a limited but meaningful community on this common interest could come into being. Separate treaties could be signed binding the signatories but reserving their freedom with respect to security preparations until *all* joined in an appropriately safeguarded disarmament program. Such a scheme would bring into a common enterprise other nations engaged in rockets and missile programs, as well as nations in other parts of the world-wide monitoring network. (Dr. Odishaw's chapter lists the countries active in these fields). It would give a sense of contribution to other countries whose contribution would be marginal technically but significant politically. Perhaps the most effective act of political symbolism within our own power would be to offer our first moon shot to the international community. A "UN Moonshot" of this nature could serve as a telling countermove against the spirit of extreme nationalism which frustrates the quest for more genuine international collaboration.

There are several activities which in international rather than national hands could result in a less unpleasant and even safer world if nations would evaluate their interests rationally. As the communications satellites mature and proliferate, we can look forward to a quantum leap in the capacity of nations to spread their propaganda. One hovering satellite in an equatorial orbit will eventually be capable of covering a sizeable fraction of the world with broadcast quality FM and even television transmission. On the more positive side, opportunities for education, literacy, and enlightenment will multiply for whole populations, if resources are available under acceptable sponsorship. Both of these functions argue, in my opinion, for the strictest possible international regulation and possibly even operation of the communication satellite system.

Even the most sensitive of our space developments—the reconnaissance satellite—might serve both the national interest and the common interest better as an internationally-operated global inspection agency with results open to all. If agreements on arms control could be reached, this step would be an obvious one. Short of that, I would like to see the United States photograph both the Soviet Union and the United States and deliver all the pictures to the UN, thus unilaterally creating an open world.

The one prediction which can be safely made is that without continuing statesmanship on the part of the Western world—preferably in collaboration with our adversaries, if necessary without them—the space age will be added to a lugubrious list that already includes the industrial revolution, the naval age, the air age, and the atomic age, all of which defied for tragically long periods of time the efforts of man to bring them within a rational framework. Only by taking prompt thought and action can we minimize the prospects of a space age also dominated by extreme nationalism and narrow national sovereignty, and thus give man a chance not to make the same mistakes all over again.

8.

Shaping a public policy
for the space age

The Difficulty of Defining Objectives

♦ JAMES R. KILLIAN, JR.

From the beginning the United States space program has been beset by the fact that public opinion has reacted to it with sharply conflicting views as to its pace, emphasis, and urgency. Let me present these contending views, first in caricature and then more soberly.

There have been unrestrained enthusiasts—derisively called the space cadets—who have viewed the space age as the gadgeteer's millenium. There have been politicians who imagined themselves being launched by rockets into a new realm of glory and power. There has been an occasional military visionary, untutored in Newtonian physics, who has exercised undisciplined imagination to conceive of weird forms of space warfare (even in the outer outer-space beyond the solar system!) offering

181

♦ James R. Killian, Jr. became Chairman of the Corporation of the Massachusetts Institute of Technology after serving eleven years as its President. From 1957 to 1959 he served as Special Assistant for Science and Technology to President Eisenhower, and President Kennedy has appointed him consultant-at-large to the President's Science Advisory Committee. Among his many consulting responsibilities, he is Chairman of the President's Foreign Intelligence Advisory Board.

his service, be it Army, Navy, or Air Force, a heaven-sent opportunity for one-upmanship in the competition for new roles and missions. Others have equated the future prestige, strength, and fate of the United States to its success in space.

In contrast to these uninhibited pro-space enthusiasts have stood the dim-viewers. To some of these the space effort—especially the man-in-space effort—has seemed parvenu and not quite respectable. To some scientists, the space enthusiasts are noisy promoters from the wrong side of the tracks. To others the space program has been a threat to their vested interests, possibly diverting funds from their pet projects. There have been anti-science types to whom space seemed the ultimate in technology run wild, a cosmic-scale boondoggle, a prodigal waste of the nation's resources, a further attack by the materialists on the already beleagured ramparts of humanism. Others, doubting the truth of the Soviet claims in space technology, have insisted that we were being misled into fruitless competition.

One needs this caricature of the extremist views in order to get a better understanding of the real issues. Many thoughtful people are apprehensive that our space program with all its drama and challenge *will* get out of hand and will absorb talent and funds to the detriment of our long-term strength in military and industrial technology and in basic science. Many feel uncomfortable about a space race with the Soviets, feeling that it is a contest which they, not we, have elected, and are not sure that in the long run this is the way to achieve maximum prestige for the nation. Still others, including many scientists and engineers, question whether the billions of dollars now being committed to space could not be better spent to meet other national needs.

Other thoughtful and reasonable people feel that we have no choice but to attempt to outdo the Soviets in this as well as other ways, that space exploration and research are a new frontier that Americans must exploit as a great and satisfying adventure, enriching the human spirit. Others argue that, even though the practical aspects of space now seem limited, when

182

measured against the great effort and funds required, it is bound to demonstrate a high level of serendipity, and thus unexpected returns in the future. Others feel that a mastery of space technology may indeed bring profound advances in military strength, and that we dare not run the risk of finding ourselves at a military disadvantage as a result of unimagined military breakthroughs made possible by space technology.

These issues and uncertainties in the minds of national leaders, members of Congress, scientists and engineers, and the general public have raised serious obstacles to the search for a definitive national space policy. Especially has it been difficult to come to a clear agreement as to the urgency and pacing of effort of our man-in-space programs, including such objectives as a manned circumnavigation of the moon or the plan to send men to the surface of the moon and to bring them back alive. I find many thoughtful people who are perplexed by these questions even though they are excited by the challenge of space exploration. The decision by the Kennedy administration to "shoot for the moon" has neither diminished the debate nor clarified the perplexity. Thus an examination of the issues is timely.

For my part of the discussion I select from the many issues only a few upon which to express my views as a concluding note to this volume. These views I know to be controversial, but I present them as part of the continuing dialogue we need in order to arrive at a clear and firm national consensus.

Some Basic Policy Considerations

PRESTIGE AS AN OBJECTIVE

Since World War II the status-seeking nations of the world community have relied increasingly on science and technology to build their prestige. The Soviets especially have used technology as an instrument of propaganda and power politics, as illustrated by their great and successful efforts— and careful political timing—in space exploration. They have sought constantly to present spectacular accomplishments in space technology as an index of national power, and too often the press and the public at large have interpreted their spectacular exploits as indices of Soviet strength and scientific accomplishment.

It must be admitted that spectacular accomplishments have temporarily enhanced the prestige of the Soviet Union, and we can all admire their achievements. But I doubt that their expensive emphasis on space exploration will be enough in the long pull to sustain either an image of strength or actual strength. This will be accomplished only by a balanced effort in science and technology. True strength and lasting prestige in science and technology will come from the richness, variety, and depth of a nation's total effort and from an outpouring of great discoveries and creative accomplishments on a wide front by its scientists and engineers. In fact, there is some reason to conclude that the over-all Soviet advance in science and technology has been retarded by their great concentration of resources and high talent on their space program. Yet one of the persistent mis-apprehensions prompted by the Soviet space feats is that Soviet science is thus revealed to be better or more advanced than United States science—although the record is convincing that American science is still superior in most respects. There has been confusion between Soviet space *technology* and science; the great Soviet achievement has been largely in the area of technology.

I believe that in space exploration, as in all other fields that we choose to go into, we must never be content to be second best; but I do not believe that this requires us to engage in a prestige race with the Soviets. We should pursue our own objectives in space science and exploration. We should neither let the Soviets choose them for us nor copy what they do. We should insist on a space program that is in balance with our other vital endeavors in science and technology and that does not rob them because they may seem to be less spectacular. In the long run we can only weaken our science and technology and lower our international prestige by frantically indulging in unnecessary competition and prestige-motivated projects.

My first point then, is that we should not misuse our science and technology by distorting them for propaganda purposes. We shall build greater respect in the long run by ensuring the quality, vigor, and integrity of all our science and technology. We shall gain prestige by being better in more areas.

COSTS AND VALUES

Closely related to the issues involved in a prestige-motivated space program are the difficult questions about pace, magnitude, and balance. It may be argued that our program to achieve a lunar landing is planned to proceed too fast and that therefore its cost will be excessive. It is feared by many, including myself, that a race with the Soviets to the moon can result in hasty engineering and the inefficient use of talent, industrial capacity, and funds. If we must have a program to achieve a successful manned round trip to the moon, I favor an orderly, thoroughly planned program, but I am not convinced that the time scale and program we are now setting for ourselves are the best way to proceed. Since I have not seen all the analyses or the detailed plans, I speak from partial knowledge, but what I do know does not reassure me that the projected costs per year from the lunar program are realistic. I am afraid that the commitments we are now making and the goals we are setting may involve expenditures far greater than the figures which have been projected. I am afraid, too, that we will actually lose time by plunging ahead too rapidly. Given these possibilities, it is of the greatest moment that we proceed with care and caution, since a space program costing billions of dollars a year will inevitably have a profound effect on the allocation of the nation's total resources of scarce technical talent and of funds, we should not permit ourselves to slide unwittingly past a point of no return without a clear understanding of the full impact of the program.

In thus urging that our plans and programs have the benefit of realistic and hard-headed study, I do not ignore what space technology can contribute to technology generally. In the same way that weapons technology has required and achieved important advances in basic technology, space technology can increase our technological dexterity and industrial capability. I am aware, too, of the possibility that the augmented space program can contribute to our economic growth, but I am not convinced that we need crash programs to put man into space to achieve this result. Crash programs, by leading to hasty execution, waste, and possible failures and delays, could reduce confidence in American technology and have adverse economic consequences.

The American people must face these questions as they seek to achieve a desirable balancing of our total national effort, particularly in the use of our scientists and engineers. If it is true that the Soviets have retarded the advance of their science and engineering in other fields by throwing too many of their very best people into their space program, then we should be alert to prevent the same effect here. We should have a space program of a magnitude and pace which we can handle well without weakening other important national programs, including defense. We must be doubly careful not to let our space program diminish the leadership we now have in the broad range of science and technology.

There is another facet of this problem of scale—the possible effects, good or bad, of "big science." In a provocative article in *Science* (21 July 1961) Alvan M. Weinberg, Director of the Oak Ridge National Laboratory, pointed up the issues sharply and indicated how complex they are:

> When history looks at the 20th century, she will see science and technology as its theme; she will find in the monuments of Big Science—the huge rockets, the high-energy accelerators, the high-flux research reactors—symbols of our time just as surely as she finds in Notre Dame a symbol of the Middle Ages. She might even see analogies between our motivations for building these tools of giant science and the motivations of the church builders and the pyramid builders. We build our monuments in the name of scientific truth, they built theirs in the name of religious truth; we use our Big Science to add

to our country's prestige, they used their churches for their cities' prestige; we build to placate what ex-President Eisenhower suggested could become a dominant scientific caste, they built to please the priests of Isis and Osiris. . . . Should we divert a larger part of our effort toward scientific issues which bear more directly on human well-being than do such Big-Science spectaculars as manned space travel and high-energy physics?

Other scientists have raised similar questions, notably Fred Hoyle, the English astronomer. He has argued against England's undertaking a large-scale space research program, maintaining that it is not worth the money and manpower and suggesting that "wherever science is fed by too much money, it becomes fat and lazy." He is worried that the tight intellectual discipline required by science is being adversely affected in the United States by "Big Science."

I do not agree with these points of view (and many scientists and engineers do not), but I respect the appropriateness of the questions which such thoughtful men ask. Their questions and comments can only leave us uneasy, especially when we face the prospect that the spirit of competition between America and the Soviet Union may lead us into even larger-scale scientific and technological undertakings, with space research and exploration coming to be one of the largest. One is uneasy too in contemplating the great areas of science and technology that need strengthening and are now neglected (e.g., materials science, oceanography) while money pours into space.

Dr. Weinberg touched upon a question of relative values that our growing space program inevitably raises. Let me state it in an oversimplified but directly blunt manner: Can we justify billions of dollars for man in space when our educational system is so inadequately supported? Does our system of values assign greater importance to this kind of exploratory activity or to the development of intellectual quality? Will an additional billion dollars a year for enhancing the quality of education not do more for the future of the United States and its position in the world than a billion more dollars a year for speeding up our program for a lunar landing?

From reading the record of Congress one can be led to conclude that a strange and hazardous inversion of values plagues our nation today. Funds to send a man to the moon seem easier to come by than funds to send a young American to a better classroom. Advances in space seem to take priority over advances in education; investment in things seems to take priority over investment in people. The same observations might be made about the urgent need for urban renewal or for adequate foreign aid.

Of course I exaggerate and over-simplify in setting up education, urban renewal, and foreign aid as competitors with the space program. But the contrast still validly lays bare the essential fact that we *do* have an acute problem of priorities, that spectacular enterprises gain support earlier than essential but less glamorous needs for the building of our human resources and for meeting human needs—and that in the end the image of America may be shaped by the quality of its inner life more than by its exploits in outer space.

THE MILITARY CONTEXT

Two other matters of program, scale, and balance deserve note at this point. One relates to the proper allocation of effort between space and defense, the other to the allocation between man-in-space projects and projects designed through instrumentation alone to collect scientific data. The collection of such data is not only important to science. It is essential to the sound planning of the continuing space program.

Given the Communist threat and until some adequately safeguarded arms limitation is achieved, unambiguous military strength is the essential base for any effective foreign policy. For the sake of the whole Free World we must not slacken in our determination to maintain military strength adequate to deter an aggressor. This is still, in my view, the surest way to deter war and to give a sense of confidence and stability to the Free World.

Thus our great military strength must be maintained and augmented. We must carry through our current program to make our nuclear striking power adequately invulnerable and thus more nearly to achieve a secure, stable deterrent. I applaud recent policy decisions to achieve an adequate limited war capability. We must seek in every way to deter brush-fire engagements and other forms of limited war. In short, an advancing military technology superior to any other in the world is essential.

Our space program should support these top priority military needs, which are certain to call for new resources of creative effort and perhaps for larger expenditures. Certainly it should not interfere with meeting them, and if properly scaled it need not. But there are dangers that it may. Let me use our ballistic missile program to illustrate this danger.

Clearly we must have ballistic missiles in such numbers, capability, and invulnerability as to give us the most powerful achievable deterrent. We need the best, the most advanced missile systems in the world. This is one race we dare not lose, and for the time being it may be more important by far than any current spare programs. Yet there is a danger that missiles are tending to become old-hat, that space projects are tending to capture the efforts of the most imaginative men and to prompt great creative effort by industry to attract new contracts. We must be alert to insure that our achievement of second- and third-generation missiles be not handicapped by a preoccupation with the "technological spectaculars" of advanced space programs. We must not permit our military strength, as it depends upon existing and on-coming weapons, to be weakened by diverting too much talent to space developments.

By the same token we should support fully those developments in space that clearly have military value such as reconaissance, communication, and meteorology. NASA has a great responsibility to support and carry through space developments important to defense, and the Department of Defense must be permitted to exploit that space technology which can authentically strengthen our military capabilities.

We should also make sure that political controversy does not prevent us from exploiting our clear lead in space communication.

MAN-IN-SPACE

Within the civilian space program there is the issue of emphasis on space science as distinct from man-in-space explorations and on the use of instruments rather than men for observation and exploration. Of course man-in-space can undertake scientific tasks of great importance, but, as experience has already demonstrated, great contributions to science can also be made through the use of instruments in space. So far our space program has been well-planned and remarkably successful in this regard; by emphasizing scientific discovery and such scientific and technological objectives as astronomy, geophysics, astrophysics, improved weather forecasting and communications, we have exploited our own special genius and proceeded in the great tradition of American science and technology. Now we are becoming more and more deeply committed to man-in-space programs. We must not let this new development diminish our emphasis on space science; we must make sure that man-in-space programs strengthen our scientific effort and do not divert us from exploiting the great power of instruments for effective exploration and observation.

Organization and Manpower

One of the issues still under debate is whether the nation is right to have a civilian space organization rather than putting all space programs under the direction of the Department of Defense. I feel strongly that our non-military space programs should be under the direction of the National Aeronautics and Space Administration and that currently we are slowly but surely finding ways to coordinate its activities with those of the Department of Defense. I was among those in 1958 who recommended a civilian space agency to President Eisenhower, and thus I may have a sense of vested interest in it; but as time has passed, I have grown more convinced that a civilian space agency is in the best interest of the United States and of its space programs. Both Keith Glennan, the first administrator, and James Webb, his successor, working closely with the Presidents and with the Secretaries of Defense involved, made good headway in resolving duplication of effort and in achieving a proper allocation of responsibilities.

I do have grave doubts about the role and function of the Space Council provided for in the Space Act. I believe it to be a fifth wheel, and that in the long run coordination and policy-making can be achieved through the normal processes of government. Clearly there will be constant problems of coordination and policy-making, but I believe they can be met by strong administrators of good will working under Presidential direction.

An agency that is going as fast as the NASA clearly is still evolving, and many problems of organization and personnel remain to be adequately dealt with. It has yet, for example, to solve satisfactorily the problem of how much work to do "in house" and how much to undertake by contract. The most impressive division of effort, and the best job of management, so far, has been in Project Mercury, which should be carefully studied as a pattern for the larger undertakings now getting underway. On the whole, the record of progress is encouraging.

One of the major requirements resting both upon NASA and the Department of Defense is to achieve the managerial organization and skills adequate to see great systems-engineering projects through to successful completion on schedule, within budgets, and with reliability. One of the rarest types of technical management talent—and one that our space programs badly need—is that which comprehends the whole array of requirements, both human and technological, which enter into the building of great engineering systems. There is a very short supply of men who can coordinate successfully the elements of research, development, test and evaluation, and production for these intricate systems. First-rate component development is not sufficient; there must be program managers who not only comprehend, with loving attention to detail, every element of the system but also can master the systems integration and the organizational and personnel problems involved in planning and administration.

When industry or government has failures in meeting performance requirements or schedules or budgets for intricate products, we can usually trace them to weakness in this kind of engineering management. As technology grows more complex and as the rate of technological obsolescence quickens, this kind of engineer-manager will come to occupy a still more crucial position in our society and certainly in our space program.

In space the need steadily grows for talent that can make decisions about technical feasibility, the kind of technical judgment and foresight that can avoid plans and requirements which lead to unnecessary engineering complexity, or programs not readily attainable within existing technology, or undertakings which will be obsolete before completed.

International Cooperation

Of the many other important issues and policy questions inherent in our space program, I can touch on only one—the furtherance of international cooperation in space research and exploration. (The great potential of space technology in the cause of peace is one of the reasons why a civilian space agency is so important in contrast to a military-dominated program.) It has been encouraging to see cooperative programs begun with England, France, and Canada. It is encouraging to see efforts getting under way by Europeans to bring European countries together for a pooled space effort. Space exploration is a "natural" for international cooperation.

Much more can be accomplished through the United Nations, and we can hope that, despite difficulties, that organization will undertake a great Conference on the Peaceful Uses of Outer Space comparable to the highly successful Conferences on the Peaceful Uses of Atomic Energy.

Despite their advocacy of international cooperation in space research, many informed American scientists have felt that it would be undesirable for the United Nations Committee on the Peaceful Uses of Outer Space to undertake space research and exploration and have strongly urged that we not support a charter for the committee that would give it such operational responsibility at this time. They were led to this position by the great success of the International Geophysical Year, which was conducted not by a political body such as the United Nations but by a private, non-political, non-government organization, the International Congress of Scientific Unions. Their conviction is that international cooperation in space research and exploration could best be encouraged and coordinated by the Space Committee (COSPAR) of this volunteer private federation. This position prevailed in the United Nations, and its Committee on the Peaceful Uses of Outer Space was limited in its responsibility to the study of the regulatory and legal aspects of space, the exchange and dissemination of information on outer space, and the encouragement of space science.

So far this seems to have been a wise position. COSPAR does not have to face political issues; and as a result, its scientist members, including the representatives of the Soviet Union, have come together in the context of a true scientific conference to reach agreements—without much more display of differences than can be expected in international scientific meetings.

Political scientists may well question—and some have—the desirability of thus by-passing an international political organization in furthering international cooperation. Is it not going to be ultimately necessary, they ask, to learn how to make the political organization effective in such matters? This is a legitimate question, but so far the scientists are supported by the unmistakable evidence that international groups of scientists seem able to achieve cooperation of great importance when they are free of political entanglements and can act freely with the tropism toward cooperation which is traditional among scientists.

Whatever may be the pattern of international cooperation, the important objective is to exploit space for peaceful purposes and to seek a space program, both within the United States and without, that is manifestly peaceful and benign. We should strive to reach agreements for space, as we leave for Antarctica, that will preclude its use for weapons of destruction.

A Point of View

In 1958 the President's Science Advisory Committee prepared a report, which President Eisenhower issued to the public, entitled "Introduction to Outer Space." That report, if I may say so, still provides an extraordinarily sound point of view toward our national space program. It gives special attention to three factors which give importance, urgency, and inevitability to the advancement of space technology.

The first of these factors is the compelling urge of man to explore and to discover, the thrust of curiosity that leads men to try to go where no one has gone before. Most of the surface of the earth has now been explored, and men now turn to the exploration of outer space as their next objective.

Second, there is the defense objective for the development of space technology. We wish to be sure that space is not used to endanger our security. If space is to be used for military purposes, we must be prepared to use it to defend ourselves.

Next, space technology affords new opportunties for scientific observation and experiment which add to our knowledge and understanding of the earth, the solar system, and the universe.

The report emphasized that our space program must take into consideration all of these objectives and the probem of national prestige as well. Its closing statements still seem valid today:

> Research in outer space affords new opportunities in science, but it does not diminish the importance of science on earth. Many of the secrets of the universe will be fathomed in laboratories on earth, and the progress of our science and technology and the welfare of the Nation require that our regular scientific programs go forward without loss of pace, in fact at an increased pace. It would not be in the national interest to exploit space science at the cost of weakening our efforts in other scientific endeavors. This need not happen if we plan our national program for space science and technology as part of a balanced national effort in all science and technology.

Our second observation is prompted by technical considerations. For the present, the rocketry and other equipment used in space technology must usually be employed at the very limit of its capacity. This means that failures of equipment and uncertainties of schedule are to be expected. It therefore appears wise to be cautious and modest in our predictions and pronouncements about future space activities —and quietly bold in our execution.

To conclude by returning to the personal views which have been the subject of this essay, let me be clear on one point. Even though I may seem to question aspects of our current space program and to emphasize many doubts about the direction we are taking, I nevertheless am convinced that space exploration is one of man's great adventures and the United States must participate in this adventure with brilliance and boldness.

Final Report of the
Twentieth American Assembly

At the close of their discussions the participants in the Twentieth American Assembly at Arden House, Harriman, New York, October 19-22, 1961, on OUTER SPACE: PROSPECTS FOR MAN AND SOCIETY, reviewed as a group the following statement. Although there was general agreement on the Final Report, it is not the practice of The American Assembly for participants to affix their signatures, and it should not be assumed that every participant necessarily subscribes to every recommendation included in the statement.

We strongly support the national space program as an intrinsically valid enterprise. Space technology promises to bring about revolutionary changes in man's life on earth. We are struck by the problem of properly allocating resources among competing national purposes, and by the opportunities offered for man to stretch his imagination and broaden his horizons. We foresee both benefits and hazards. We welcome the challenge afforded the United States by the space program to advance science, to improve international cooperation, and to utilize its leadership for the general welfare of all mankind.

Our capacity to anticipate constructively the consequences of the space age is already being tested. Satellite communication systems will soon gird the globe. Weather prediction has even now crossed the barrier hitherto imposed by earthbound observations. Man has left the earth's atmosphere and circled the world in little over an hour, living to tell the story. As yet we have seen but the beginning of a revolution the forces and opportunities of which will grow at an accelerating rate.

The Soviet Union has led the way in a number of concrete space achievements. We believe that the United States leads the way in the space sciences and their peaceful applications, and we are confident that the United States' major space objectives will be accomplished. Both Soviet and United States efforts can be beneficial to mankind.

But like atomic energy, which also holds great potential for human well-being, space technology has two faces. The negative aspect of space is its potential for human destruction. In an era of persistent hostility from the expansionist forces of world communism, all weapons and all technologies become involved in the Cold War. We are concerned that outer space, like the other dimensions—earth, sea, and air—can be employed by the forces dedicated to the overthrow of free societies to threaten their security.

These considerations all pose for the American people and their leaders basic issues of national policy.

Priorities and Resources

The space program ranks with the most pressing American undertakings and carries a high priority. The size, intensity and direction of our present effort have been shaped largely by psychological and strategic considerations. But the national effort in space has other major objectives in the realms of science, civil applications, and international cooperation, which would be valid regardless of the Soviet program.

The United States has the capacity to support its space activities and at the same time to maintain a proper balance with other programs in our national interest. We should become accustomed to sustaining the space programs operated by the National Aeronautics and Space Administration (NASA) over the coming years. For example, the budget next year will be on the order of $3.5 billion and may be higher in succeeding years. There will also be a continuing requirement for support of a substantial space program in the Department of Defense. We are pleased that our space activities have wide bi-partisan support in the Congress.

Space programs might create shortages of manpower at the highest level, and we should make plans to avoid depriving other areas of leadership. The space program should be regarded as giving the nation a new incentive further to improve the level and quality of American scientific and technical education.

We support the enthusiasm and vigor with which the government has defined its space goals, including the manned lunar landing. But we caution against excessive focus on that project as either a definitive test of relative Soviet and American capabilities, or as a scientific or technical end in itself. The lunar project is one major step in a continuing program for the scientific exploration of the solar system. However, both the American position in the world and the over-all relationship between East and West clearly rest on far broader considerations.

The Need to Plan

Just as space technology has its own "lead times," so better methods must be developed to anticipate and plan for the political, social and economic consequences of space technology. Both the legislative and executive branches of the federal government and appropriate industries are to be commended for the preliminary planning they have done, particularly with respect to communication satellites. But far more attention needs to be given to planning by the government, by business, and by universities and other research organizations.

Organizing to Act

The space program presents continuing problems of organization and management. We are not convinced that the difficulties inherent in a complex enterprise such as space can be resolved by formal organizational structure.

Since space science has both military and civil applications, there will be continuing overlap between the two. On balance, the decision favoring civil emphasis in administration of our national space program is sound; it neither impinges on nor reduces the importance of our military program. However, further improvement in civil-military liaison is essential at both the working and policy levels. We recommend that the administrative structure underlying the organization of federal space activities be kept under continuous review.

In playing their role in research, development and production of space systems, American business and industry will continue to operate under government direction and, for the most part, have financial support from the government. A wide variety of contractual arrangements has been devised, under which private corporations assist in the execution of public policy. The government should seek under such arrangements to make maximum use of private institutions to advance the public interest.

Communication Satellites

We emphasize the desirability of establishing a practical system of communication satellites at the earliest possible date. Such a system has enormous potential for expansion and improvement in international communications and the fostering of international understanding. We support our government's policy which favors private operations provided they meet the goals and standards set forth in that policy. We believe that our objective should be a communication satellite system of global coverage with provision for participation by other countries and international agencies.

Cooperation Among Scientists

We believe it highly desirable to enlarge fruitful professional contacts through non-government scientific groups, while recognizing the differences in degree of political constraint imposed on participants from different societies. We hold the advancement of scientific knowledge to be a worthy end in itself, without expecting definitive effects on the larger political and strategic confrontation.

We commend NASA's program of making provision in the payloads of United States rockets for scientific experiments by interested countries. The United States should press vigorously for holding the proposed international scientific conference on outer space recommended by the United Nations General Assembly. The flow of ideas and scientific data in such a forum, as well as the opportunity to demonstrate our own advances, will be of considerable benefit to the United States as well as to the cause of international cooperation.

Arms and Arms Control

Capabilities in outer space have become an important reflection of national power, particularly as the threat of weapons is increasingly employed by the Soviet Union for political ends. To meet this threat the United States must strengthen research and development in support of an effective military space capability.

Our government should make additional efforts to seek safeguarded agreements banning weapons of mass destruction in space before they become accomplished facts. Mutual pre-launching inspection of rockets could be done independently of other arms controls.

Short of a genuine disarmament agreement, the search needs to be expanded for areas of agreement designed to enhance the stability of the military environment. To that end we recommend accelerated studies inside and outside government on the technical and political problems of arms control in space.

International Control

There is already an international consensus that outer space is freely available for use by all nations in accordance with international law, and a consensus is developing that celestial bodies are incapable of appropriation to national sovereignty. We urge our government to seek agreement at the United Nations and elsewhere on these and other basic principles.

We urge that the United Nations and its specialized agencies be given increased responsibilities with respect to international communication and weather satellite systems. We further recommend the creation within the

United Nations Secretariat of appropriate staff to perform necessary co-ordinating functions in the field of outer space. Bilateral and other international joint action and programs should also be encouraged where necessary to advance the primary goal of fostering the rapid development of rational and efficient operations in outer space activities.

We believe it would be highly desirable for the United Nations Committee on the Peaceful Uses of Outer Space to get on with its job. The United States government should intensify its effort to achieve this end, but without accepting the Soviet veto formulas which compromise traditional principles as well as the normal rules of procedure governing United Nations committee operations.

*   *   *   *

Before long, human footsteps will imprint the dust of the moon's surface. As more and more satellites cross the skies, and as man's old dream of contact with the moon and the planets is transformed into reality, human life will inescapably take on new dimensions. Not the least of the motives impelling us will be the human and cultural values involved in pursuing the high goals of knowledge about our origins and our destiny. Few are the generations privileged to take part in a comparable enterprise. For this adventure, and for all of its social, economic, scientific, military and political consequences, our nation can and should pledge both its generous support and its responsible leadership.

Participants in the Twentieth American Assembly

ROY ALEXANDER
Editor, *Time* Magazine
New York City

ARTHUR G. ALTSCHUL
Partner
Goldman, Sachs & Co.
New York City

JAMES P. BAXTER, III
Council on Foreign Relations
New York City

CLAUDE BLAIR
Vice President
American Telephone & Telegraph Company
New York City

JOHN BOETTIGER
Harriman Scholar
Columbia University

HUGH BORTON
President
Haverford College
Pennsylvania

THOMAS C. BOSTIC
President, Cascade Broadcasting Co. &
Mayor of Yakima, Washington

IAN D. BOYD
Harriman Scholar
Columbia University

DONALD G. BRENNAN
Lincoln Laboratories
Massachusetts Institute of Technology

DAVID A. BURCHINAL
Major General
United States Air Force
Washington, D. C.

LOUIS W. CABOT
President
Godfrey L. Cabot, Inc.
Boston

WARD M. CANADAY
President
The Overland Corporation
Toledo

E. FINLEY CARTER
President
Stanford Research Institute
California

BOB CASEY
Representative from Texas
Congress of the United States

EMILIO G. COLLADO
Director, Standard Oil Co. (N. J.)
New York City

J. RUSSELL CLARK
Vice President and General Manager
  (Astronautics)
Chance Vought Corporation
Dallas

PHILIP K. CROWE
Former United States Ambassador to
  Ceylon
Easton, Maryland

EMILIO Q. DADDARIO
Representative from Connecticut
Congress of the United States

LINCOLN P. BLOOMFIELD
Center for International Studies
Massachusetts Institute of Technology

KARL DEUTSCH
Professor of Political Science
Yale University

GOLDTHWAITE H. DORR
Dorr, Hand, Whittaker & Watson
New York City

WILLIAM H. EDWARDS
Edwards & Angell
Providence, Rhode Island

OSBORN ELLIOTT
Editor, *Newsweek* Magazine
New York City

PHILIP J. FARLEY
Special Assistant to the Secretary of
  State
Washington, D. C.

JAMES G. FULTON
Representative from Pennsylvania
Congress of the United States

EILENE GALLOWAY
U. S. Senate Committee on Aeronautical
  and Space Sciences
Washington, D. C.

RICHARD N. GARDNER
Deputy Assistant Secretary of State
Washington, D. C.

JOSEPH M. GOLDSEN
RAND Corporation
Santa Monica, California

STANLEY T. GORDON
Ford Foundation
New York City

H. FIELD HAVILAND, JR.
The Brookings Institution
Washington, D. C.

WILLIAM P. HOBBY, JR.
Managing Editor
The *Houston Post*
Texas

ROBERT JASTROW
Director, Institute for Space Studies
New York City

JOHN JOHNSON
General Counsel
National Aeronautics and Space Adminis-
  tration
Washington, D. C.

AMROM H. KATZ
RAND Corporation
Santa Monica, California

ALLAN KLINE
Western Springs, Illinois

DEXTER MERRIAM KEEZER
Economic Advisor
McGraw-Hill Publishing Company
New York City

ALAN G. KIRK
Admiral, U.S.N. (Ret.)
New York City

G. A. LINCOLN
Colonel, U.S.A.
Head, Department of Social Sciences
United States Military Academy
West Point

LEON LIPSON
Professor of Law
Yale University

LEONARD C. MEEKER
Legal Advisor, Department of State
Washington, D. C.

ROBERT E. MERRIAM
President
Spaceonics, Inc.
Geneva, Illinois

DONALD N. MICHAEL
The Peace Research Institute
Washington, D. C.

JOHN S. MILLIS
President
Western Reserve University

SIR LESLIE MUNRO *
Secretary General
International Commission of Jurists

J. MORDEN MURPHY
Vice President (Ret.)
Bankers Trust Company
New York City

CALVIN NICHOLS
Director, World Affairs Council of Northern California
San Francisco

MARTIN S. OCHS
Editor, *The Chattanooga Times*
Tennessee

HUGH ODISHAW
Executive Director
Space Science Board
National Academy of Sciences
National Research Council
Washington, D. C.

JACK C. OPPENHEIMER
Office of Plans & Program Evaluation
National Aeronautics & Space Administration
Washington, D. C.

———
* Delivered formal address.

DAVID PERLMAN
Science Editor
San Francisco Chronicle
California

FRANCIS T. P. PLIMPTON
The Deputy Representative of the United States to the United Nations

R. W. PORTER
Consultant—Advanced Development
General Electric Company
New York City

WESLEY W. POSVAR
Colonel, U.S.A.F.
Head, Department of Political Science
United States Air Force Academy
Colorado

DON K. PRICE
Dean, Graduate School of Public Administration
Harvard University

ROSS J. PRITCHARD
Chairman, Department of International Studies
Southwestern at Memphis
Tennessee

RICHARD RHOADS
Harriman Scholar
Columbia University

MICHAEL ROSS
Director, Department of International Affairs
AFL-CIO
Washington, D. C.

BERNARD A. SCHRIEVER
Lieutenant General, U.S.A.F.
Air Force Systems Command
Andrews Air Force Base
Washington, D. C.

MARSHALL D. SHULMAN
The Fletcher School of Law and Diplomacy
Tufts University

LEONARD SILK
Senior Editor, *Business Week*
New York City

HOWARD J. TAUBENFELD
Professor of Law
Southern Methodist University

JAMES E. WEBB *
Administrator
National Aeronautics & Space Administration
Washington, D. C.

FRANCIS O. WILCOX
Dean, School of Advanced International Studies
The Johns Hopkins University

CARROLL L. WILSON
School of Industrial Management
Massachusetts Institute of Technology

ROBERT J. WOOD
Lieutenant General
United States Army Air Defense Command
Colorado Springs

CHRISTOPHER WRIGHT
Executive Director
Council for Atomic Age Studies
New York City

GEORGE W. WRISTON
Spencer, Trask & Co.
Albany

* Delivered formal address.

The American Assembly

A merican Assembly books are purchased and put to use by thousands of individuals, libraries, businesses, public agencies, non-governmental organizations, educational institutions, discussion meetings and service groups. *The subjects of Assembly studies to date are:*

1961—ARMS CONTROL
—OUTER SPACE

1960—THE SECRETARY OF STATE
—THE FEDERAL GOVERNMENT AND HIGHER EDUCATION

Library, cloth bound edition, $3.50
Spectrum, paper bound edition, $1.95
Available from better booksellers and Prentice-Hall, Inc.

The following titles were published by The American Assembly. Prices indicate books which can be obtained by writing to The American Assembly.

1959—THE UNITED STATES AND LATIN AMERICA ($2.00)
—WAGES, PRICES, PROFITS AND PRODUCTIVITY ($2.00)

1958—THE UNITED STATES AND AFRICA ($2.00)
—UNITED STATES MONETARY POLICY ($2.00)

1957—ATOMS FOR POWER
—INTERNATIONAL STABILITY AND PROGRESS

1956—THE UNITED STATES AND THE FAR EAST
—THE REPRESENTATION OF THE UNITED STATES ABROAD

1955—THE FORTY-EIGHT STATES
—UNITED STATES AGRICULTURE

1954—THE FEDERAL GOVERNMENT SERVICE
—THE UNITED STATES STAKE IN THE UNITED NATIONS

1953—ECONOMIC SECURITY FOR AMERCANS

1952—INFLATION

1951—UNITED STATES-WESTERN EUROPE RELATIONSHIPS

Regular readers of The American Assembly receive early copies of each new Assembly study and are billed subsequently. Future Assembly books, to be published by Prentice-Hall, Inc., are:

—AUTOMATION
—INTERNATIONAL EDUCATIONAL, CULTURAL, AND SCIENTIFIC ISSUES FOR THE UNITED STATES IN THE 1960'S

To enroll as a regular reader, or for additional information, please address:

The American Assembly, Columbia University, New York 27, New York.

SPECTRUM  PAPERBACKS

*Other SPECTRUM Books . . . quality paperbacks that
meet the highest standards of scholarship and integrity.*

S-31 HEROES, VILLAINS, AND FOOLS: THE CHANGING AMERICAN CHARACTER, Orrin E. Klapp, *paper* $1.95, *cloth* $3.95

S-32 COMMUNIST CHINA'S STRATEGY IN THE NUCLEAR ERA, Alice Langley Hsieh, *paper* $2.25, *cloth* $4.50

The American Assembly Series

S-AA-1 THE FEDERAL GOVERNMENT AND HIGHER EDUCATION, edited by Douglas M. Knight, *paper* $1.95, *cloth* $3.50

S-AA-2 THE SECRETARY OF STATE, edited by Don K. Price, *paper* $1.95, *cloth* $3.50

S-AA-3 GOALS FOR AMERICANS, THE REPORT OF THE PRESIDENT'S COMMISSION ON NATIONAL GOALS, *paper* $1.00, *cloth* $3.50

S-AA-4 ARMS CONTROL ISSUES FOR THE PUBLIC, edited by Louis Henkin, *paper* $1.95, *cloth* $3.50

S-AA-5 OUTER SPACE, edited by Lincoln P. Bloomfield, *paper* $1.95, *cloth* $3.95

S-AA-6 THE UNITED STATES AND THE FAR EAST (Second Edition), edited by Willard L. Thorp, *paper* $1.95, *cloth* $3.95

Science and Technology Series

S-ST-1 THE ATOM AND ITS NUCLEUS, George Gamow, *paper* $1.95, *cloth* $3.75

S-ST-2 ROCKET DEVELOPMENT, Robert H. Goddard, *paper* $2.45, *cloth* $3.95

Classics in History Series

Twentieth-Century Views Series